A Single Layer Revolution: Unveiling the Promise of Graphene

Shabnam

Contents

Introduction

The breakthrough in the research of atomically thin materials took place in 2004 with the experimental isolation of a single-layer of graphite, denominated graphene [1]. This experimental success was possible due to the layered crystal structure of the graphite. The atomic structure of bulk graphite is composed by layers of carbon atoms arranged in a hexagonal honeycomb lattice. The layers are weakly bound to each other. In fact, the interaction between layers is mostly due to van der Waals forces with very weak covalent bonds. For this reason, the isolation of graphene was relatively easy and possible by simple mechanical cleavage.

The discovery of graphene very quickly attracted the attention of the scientific community. It is indeed a very exotic system that was never available before. Besides the intuitive beauty and interest of this system, graphene showed promising electronic and mechanical properties for technology. The first electronic properties that have to be mentioned are the vanishing band gap in the K-valley and the linear dispersion of the band structure in the vicinity of the intersection of the valence and conduction bands [2]. These very peculiar properties triggered many fascinating transport and optical experiments, e.g. anomalous quantum Hall effect and Landau levels [3] or the optical imaging of plasmons [4], and the development of promising technology, e.g. broadband optical modulators or photo-detectors [5].

Furthermore, graphene has excellent mechanical properties due to its large elasticity. This is why one of the main potential application of graphene is flexible electronics [6].

Because of these intriguing discoveries, the scientific community started to isolate single layers of other layered bulk crystals. The research in this field proceeded very swiftly and today there is a huge variety of atomically thin materials. This set of materials is often called van der Waals materials or two-dimensional materials. The research on these materials was further motivated by the possibility to stack on top of each other different two-dimensional materials via van der Waals interaction [7]. Because of the nature of the latter interaction, the lattice mismatch between two different compounds does not limit the fabrication of heterostructures. This gives the freedom to combine all kind of materials and to create atomically thin metamaterials.

The scope of this **book** is to investigate the optical and infrared properties of atomically thin semiconductors, with a special focus on one of the most promising subsets of two-dimensional semiconductors: transition metal dichalcogenide (TMD) monolayers. Differently from graphene, TMDs have a finite direct optical band gap in the monolayer limit. This feature, together with their unique two-dimensional lattice structure makes this class of materials very appealing.

Various experiments that have been performed on these materials are presented in this **book**. The most relevant results are presented in Chapter 3 and in Chapter 4. The main topic of Chapter 3 is the localization of excitons by adsorbed gas molecules in a $MoSe_2$ monolayer. Other details reported are for instance the effects of laser irradiation on the localization and the role of exciton hopping between localization sites. In Chapter 4, different experiments performed with the infrared free-electron laser are discussed. The main focus in this chapter is on the observation of the redshift of the trion resonance induced by infrared radiation. A detailed theoretical description of this process is also presented.

But besides TMDs, the 2D world is much larger and more composite. Other semiconductor compounds were studied during the course of this PhD and they will be partly presented in this **book**. In particular, the work on InSe crystals will be presented in Chapter 5. Since InSe degrades under ambient conditions, the flakes were encapsulated in hexagonal boron nitride (hBN). The effects of the encapsulation will be described, focusing mostly on the optical properties. Furthermore, the dynamics of the photo-excited electron-hole system will be discussed, especially when the thickness of the InSe crystals decreases up to few nanometers.

Finally, TMD semiconductor heterostructures were fabricated and investigated during the **book**. The preliminary results on this topic will be presented in the last Chapter 6, from the fabrication of the heterostructures to the optical characterization and the first experimental results.

1. Atomically thin semiconductors

This chapter provides an introduction for the research presented in this book. The main properties of all the materials investigated in the experiments are out-lined. Furthermore, the motivation and the state-of-the-art of the research in each field is discussed. The chapter is divided in three sections and is organized as follows:

- After an introduction about the physical properties and about the state-of-the-art of the research on TMD monolayers, the optical properties are described in detail with a special attention to the exciton physics and the valley degree of freedom. A general theoretical description of excitons in semiconductors is given as well.

- An overview of the research on two-dimensional semiconductors is presented. Afterwards, a section is dedicated to describe the physical and optical properties of InSe.

- An introduction on van der Waals heterostructures is presented, mostly focusing on TMD semiconductor heterostructures.

1.1. Transition metal dichalcogenide monolayers

Transition metal dichalcogenide monolayers are single layers of crystals that have a transition metal atom (Mo, W, Ti, Hf) and two chalcogenide atoms (S, Se, Te) in the primitive cell. Most of these compounds crystallize in a honeycomb structure (2H-phase), as shown in Figure 1.1 (top view). The bulk crystal is composed of many layers weakly bound to each other via van der Waals forces. For this reason, the isolation of a monolayer is relatively easy and, in most cases,

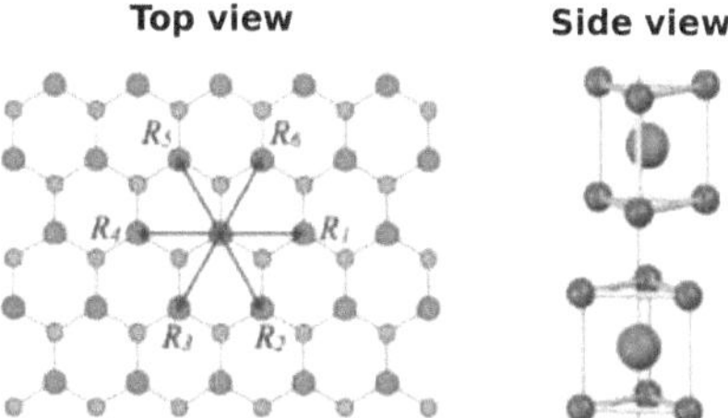

Figure 1.1.: Top and side view of the honeycomb lattice (trigonal prismatic) of TMD crystals in the 2H-phase. The image is adjusted from [8].

a single layer is stable. The vertical alignment of the monolayers in the bulk crystal is shown in Figure 1.1 (side view).

TMD crystals have different electronic properties, from insulators and semiconductors ($HfSe_2$ or MoS_2) to metals (VS_2 or NbS_2), depending on the compound and on the crystal phase. Furthermore, some compounds have peculiar properties such as $MoTe_2$ and WTe_2 which are Weyl semimetals in the orthorhombic phase [9]. In this **book**, only semiconductor compounds were studied. In particular, MoS_2, $MoSe_2$, and WSe_2 in the 2H-phase were investigated. In the following sections, only the properties of these latter compounds will be discussed.

These materials are the most studied transition metal dichalcogenide monolayers. Remarkably, they are stable in air under ambient condition without any significant degradation on a timescale of months or years [10]. This is not a common feature among two-dimensional materials. In fact, the extreme surface-to-volume ratio makes this class of materials very sensitive to air exposure and different kind of degradation processes.

1.1.1. Optical properties: excitons

The electronic and optical properties of semiconducting TMDs are strongly dependent on the number of layers. The first striking behavior is the indirect-to-direct band-gap crossover when the material reaches the monolayer limit [11, 12]. In fact, as shown for the first time in 2010, a monolayer of TMD has a direct band gap, while crystals with more than one layer have indirect band gap. A sketch of the evolution of the band structure with number of layer is depicted in Figure

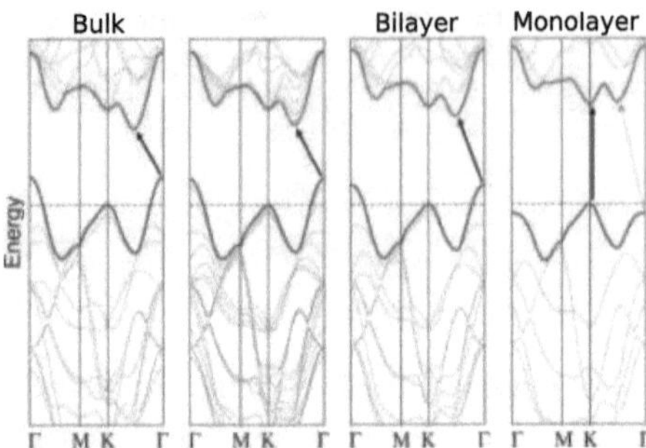

Figure 1.2.: Layer dependence of the band structure of MoS_2 calculated with Density Functional Theory (DFT). Bulk MoS_2 has an indirect band gap. Only in the monolayer limit, the band gap becomes direct. The graph was taken by [11].

1.2.

This fact immediately attracted the attention of the scientific community. A direct band-gap semiconductor thick as an atom would be extremely beneficial for a lot of semiconductor technology. Prototypes of various opto-electronic devices have already been fabricated and tested, as for instance optical detectors, solar cells, and lasers based on monolayer MoS_2 [13–16].

The energy of the optical band gap of TMD monolayers depends on the compound and it ranges from the near-infrared to the visible. Figure 1.3 shows photoluminescence (PL) spectra at room temperature of three different TMD compounds.

However, according to DFT numerical calculations, the energy band gap of these materials should be at higher energies in comparison with the optical band gap observed in PL. The difference is around 0.5 eV and it varies slightly for different materials. The origin of this energy difference is the exciton renormalization of the optical band gap. An exciton is an electron-hole pair in a semiconductor bound by the Coulomb attraction. The attractive Coulomb interaction reduces the actual energy of the optical band gap.

Typical exciton binding energies are on the order of $3 - 30$ meV for bulk semiconductors and quantum wells. TMD monolayers have exciton binding energies on the order of hundreds of meV. This high binding energy means that the inter-

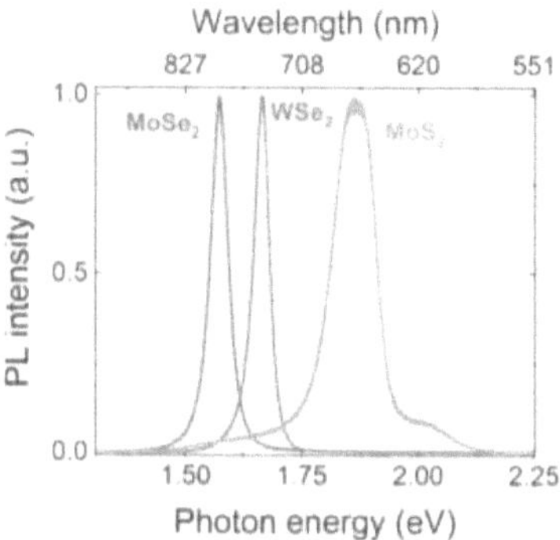

Figure 1.3.: PL spectra of MoS$_2$, WSe$_2$, and MoSe$_2$ monolayers.

action between electron and hole is extremely strong. The reason is the missing dielectric screening of the lattice environment because of the atomic thickness of the vdW materials.

Direct measurements of the exciton binding energy were performed with different experimental techniques. One way to measure the exciton binding energy is by measuring the series of the excited states of the exciton (Figure 1.4). In a simplified picture, an exciton can be modeled as a hydrogenic atom. This model leads to an infinite number of solutions that are the internal quantum states of an exciton. The eigenvalues of the bound exciton states in two dimensions are:

$$E_X^{(n)} = \frac{m^* e^4}{2\hbar^2 \epsilon^2 (n - \frac{1}{2})^2},$$
(1.1)

where $E_X^{(n)}$ is the energy of the n-th exciton state, m^* is the reduced exciton mass, ϵ is the dielectric constant, and n is the exciton principal quantum number. These resonances can be observed with linear spectroscopy, as shown in Figure 1.4. In this latter example, the reflectivity spectrum is taken from a WS$_2$ monolayer. An exciton binding energy 320 meV was extracted from the experiment. Other TMD monolayer compounds have similar exciton binding energy values [17–19].

Equation (1.1) is derived by the solution of the Wannier equation of an exciton confined in a plane but with an interaction that is able to propagate in a three-dimensional space. For an exciton in a three-dimensional space, the solution is

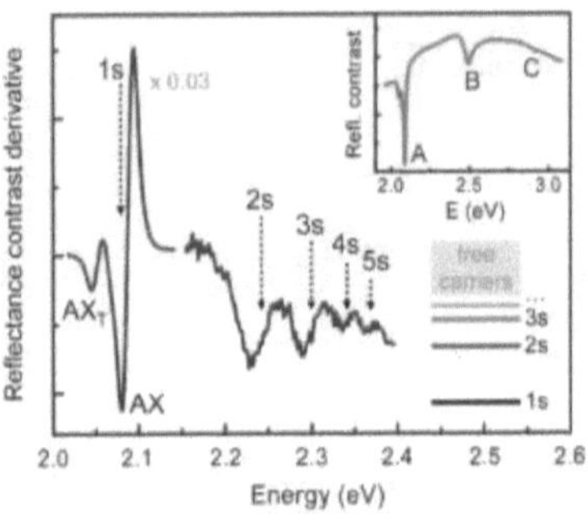

Figure 1.4.: Reflectance derivative of WSe$_2$ monolayer at low temperature. The sequence of the internal exciton states is clearly observable. The image was taken from [20].

still the same with the substitution $n - \frac{1}{2} \longrightarrow n$. In this last case, the more familiar sequence of solutions of the electron states of a hydrogen atom is retrieved. The term $-\frac{1}{2}$ has part of the responsibility for the large exciton binding energy in TMD monolayers.

Since the binding energy is so high in this class of materials, excitons are stable even at room temperature. Technology based on excitons at room temperature has already been experimentally proved in TMD monolayers, as for instance an excitonic laser using MoS$_2$ monolayer as gain medium [15].

Because of the huge exciton binding energy in TMD monolayers, excitons dominate the optical response of these materials and they can never be neglected. Since exciton physics is one of the main topics in this **book**, a mathematical introduction to excitons will be given in the next section, starting from the general Semiconductor Bloch Equations.

1.1.2. Theoretical description of excitons

The Semiconductor Bloch Equations (SBE) are a set of differential equations that describe the optical response of a semiconductor under the excitation by an electromagnetic field. The treatment is very general and, by means of these equations, it is possible to describe the physics behind many topics of contemporary research, as polaritons, excitons, many-body complexes, and their dynamics.

The equations are derived under the assumption of a two-band system - valence and conduction bands. In this regard, the treatment is purely quantum mechanical. However, the electromagnetic field follows the classical treatment. More sophisticated approaches for the treatment of exciton are possible, for instance including the Maxwell equations for the electromagnetic field.

The semiconductor Bloch equations describe the dynamics of the semiconductor system following the evolution of the microscopic polarization P_k and the electron and hole population $f_{e,k}$ and $f_{h,k}$, respectively, where k is the wavevector. The SBEs are:

$$i\hbar\frac{\partial P_k}{\partial t} = \epsilon_k - (1 - f_{h,k} - f_{e,k})\omega_{R,k} + i\hbar\left.\frac{\partial P_k}{\partial t}\right|_{scat}, \tag{1.2}$$

$$\hbar\frac{\partial f_{e,k}}{\partial t} = -2\operatorname{Im}(\omega_{R,k}P_k^*) + \hbar\left.\frac{\partial f_{e,k}}{\partial t}\right|_{scat}, \tag{1.3}$$

$$\hbar\frac{\partial f_{h,k}}{\partial t} = -2\operatorname{Im}(\omega_{R,k}P_k^*) + \hbar\left.\frac{\partial f_{h,k}}{\partial t}\right|_{scat}, \tag{1.4}$$

where ϵ_k is the renormalized energy of the electron-hole pair and ω_R is the renormalized Rabi frequency [21]. The renormalized energies and Rabi frequency are:

$$\epsilon_k = \epsilon_k^c - \epsilon_k^v + \sum_{k\neq q} V_{k-q}P_q$$

$$\omega_R = \frac{1}{\hbar}\left(\mathbf{d}\cdot\mathbf{E}(t) + \sum_{k\neq q} V_{k-q}(n_{e,k} + n_{h,q})\right),$$

where V_k is the Coulomb interaction, $\mathbf{d}$ is the dipole matrix moment of the interband transition, and $\mathbf{E}$ is the external electromagnetic field. The renormalization takes into account the electric field of all the local dipoles in the semiconductor. This is basically a mean field that each electron and hole with a certain momentum feels.

The SBEs in Eq. 1.2 - 1.4 are exact. All the high-order scattering processes as electron-phonon, electron-electron scattering and so on are included in the scattering part of the polarization and electron and hole distributions.

A full derivation of the SBEs can be found in many textbooks, for instance here [22].

Going from the electron-hole to the conduction-valence band notation, i.e.

$$n_v = 1 - f_v,$$
$$n_c = f_c,$$

the SB equation for the polarization becomes:

$$i\hbar\frac{\partial P_k}{\partial t} = (e_c - e_v)P_k + (n_{c,k} - n_{v,k})\omega_{R,k} + i\hbar\left.\frac{\partial P_k}{\partial t}\right|_{scat}, \tag{1.5}$$

where e_c and e_v includes the renormalization term for each band.

In order to get an equation for excitons, the SBEs are considered neglecting the scattering terms. This corresponds to a random phase approximation (RPA), i.e. only the interaction terms with the same wavevectors are taken into account. Furthermore, the formation of many-body particle as trions or biexcitons is neglected. In fact, these terms come from the three- and four-particle interactions contained in the scattering term. A derivation of the trion microscopic polarization is discussed in Appendix A.4. The equation for the microscopic polarization becomes:

$$\left(i\hbar\frac{d}{dt} - (e_{c,k} - e_{v,k})\right)P_k(t) = \hbar(n_{c,k} - n_{v,k})\omega_{R,k}. \tag{1.6}$$

This equation gives the dynamics of interband microscopic polarization depending on the electron-hole distribution in the valence and conduction bands.

The terms:

$$\mathrm{Im}(\omega_{R,k}P_k^*) = \mathrm{Im}\left(\frac{1}{\hbar}\left(\mathbf{d}\cdot\mathbf{E}(t) + \sum_{k\neq q}V_{k-q}(n_{e,k} + n_{h,q})\right)P_k^*\right) \tag{1.7}$$

in the second and third SBEs describe the generation of electrons and holes after absorption. In the RPA, i.e. without any scattering, the generation rates for holes and electrons are identical and stay identical in time, as it should be.

Excitons are the bound state solutions of Eq. (1.6). However, excitons exist regardless of their temporal evolution. Therefore, it is possible to make the assumption of quasi-thermal equilibrium. In this case, n_c and n_v approach the limit of Fermi-Dirac distributions, i.e. $n_{c,k}(t) = f_{c,k}$ and $n_{v,k}(t) = f_{v,k}$. Eq. (1.6) becomes:

$$\left(i\hbar\frac{d}{dt} - (e_{c,k} - e_{v,k})\right)P_k(t) = (f_{c,k} - f_{v,k})\left(\mathbf{d}\cdot\mathbf{E}(t) + \sum_{k\neq q} V_{k-q}P_q\right). \qquad (1.8)$$

The interaction term $\sum_{k\neq q} V_{k-q}P_q$ is responsible for the coupling between electrons and holes or, in other words, for the formation of excitons. In the limit of vanishing interaction, it is possible to solve analytically the equation above and to get the free particle absorption.

The term $f_{c,k} - f_{v,k}$ is the phase space filling or Pauli blocking factor. This factor takes into account the excited population of electrons. For simplicity, $f_c = 0$ and $f_v = 1$ are assumed for the rest of the discussion.

Since $e_{c,k} - e_{v,k} = E_g + \frac{\hbar^2 k^2}{2\mu^2}$, where $\mu = \left(\frac{1}{m_v} + \frac{1}{m_c}\right)^{-1}$ is the reduced mass, one gets:

$$\left(i\hbar\frac{d}{dt} - E_g - \frac{\hbar^2 k^2}{2\mu^2}\right)P_k(t) = -\left(\mathbf{d}\cdot\mathbf{E}(t) + \sum_{k\neq q} V_{k-q}P_q\right). \qquad (1.9)$$

Expressing the last equation in the real space via Fourier transformation, one gets:

$$\left(\hbar(\omega + i\delta) - E_g + \frac{\hbar^2\nabla^2}{2\mu^2} + V(r)\right)P(r,\omega) = -\mathbf{d}\cdot\mathbf{E}(\omega)\delta_r. \qquad (1.10)$$

The solutions of this equation give the optical response of a semiconductor in the presence of an external electromagnetic field.

The homogeneous part of Eq. (1.10) is:

$$-\left(\frac{\hbar^2\nabla_\mathbf{r}^2}{2\mu^2} + V(r)\right)\psi_n(\mathbf{r}) = E_n\psi_n(\mathbf{r}), \qquad (1.11)$$

where $E_n \equiv E'_n + E_g$ and E'_n are the eigenvalues of Equation 1.11.

This equation is called Wannier equation and has the same form of a two-particle Schroedinger equation, i.e. the hydrogen problem. Solving this equation is a very well known problem and analytically possible. As for the hydrogen atom, there are two classes of solutions, one for $E_n < E_g$ and the other for $E_n > E_g$. The first class gives an infinite and convergent sequence of solutions, while the second a continuum of states.

The physical interpretation is that for $E_n < E_g$ the interacting electron and hole are in a common bound state and they form a stable pair. This pair is called

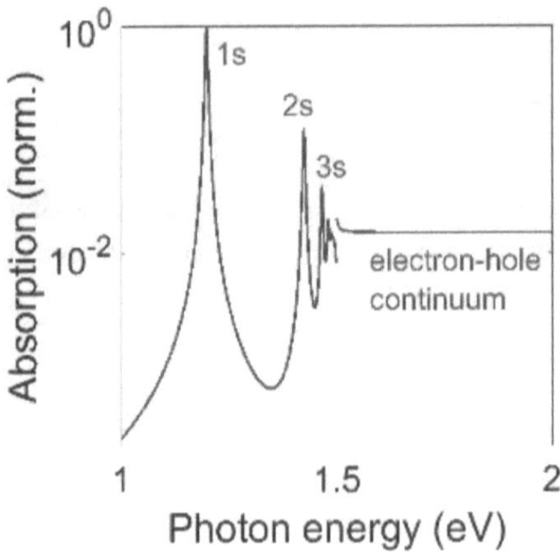

Figure 1.5.: Absorption spectrum of a direct band-gap semiconductor with 1.5 eV band gap and 0.3 eV exciton binding energy. The spectrum was calculated using Elliott's formula (Appendix, Eq. A.5).

exciton. Differently, above a certain energy E_g, the electron-hole pair is not stable and bound anymore and they behave as free electrons and holes.

From Eq. (1.10), it is possible to derive Elliott formulas for absorption and expressions for photoluminescence (a short derivation is reported in appendix A.1). An example of absorption spectrum is calculated in Figure 1.5. The exciton series is clearly visible at energies below the free electron-hole band gap.

In the absorption spectrum, there is a discontinuity at the electron-hole band gap. This can be removed introducing an inhomogeneous or thermal broadening. A similar approach was followed in the experiment presented in chapter (5) in order to get an analytical expression for photoluminescence spectra.

1.1.3. Valley degree of freedom

This section is dedicated to another striking feature of excitons in TMD monolayers, that is the valley-dependent polarization of excitons.

The mathematical treatment will be discussed in the free electron-hole picture but the conclusions hold when the formation of exciton pairs via Coulomb interaction is included.

As shown in the band structure in Figure 1.2, TMD monolayers have a direct band gap at the K point of the Brillouin zone. Due to the symmetry of the system, the direct band gap is six-fold degenerate in bulk TMD. However, in the monolayer limit, one symmetry is lost, i.e. the inversion symmetry. This is the inversion symmetry breaking.

Mathematically, the group symmetry of the crystal changes from D_{6h}^4 to D_{3h}^1 [23]. This means that in the reciprocal space, K and K' valleys are not anymore identical. This leads to different optical selection rules for the polarization in the two valleys.

In fact, in the new symmetry group, the wave functions at the K(K')-points in the valence and conduction bands are respectively:

$$|\psi_v^\tau\rangle = \frac{1}{\sqrt{2}}(|d_{x^2-y^2}\rangle + i\tau\,|d_{xy}\rangle),\tag{1.12}$$

$$|\psi_c^\tau\rangle = d_{z^2},\tag{1.13}$$

where $d_{x,y,z}$ are the basis functions of the conduction and valence bands (d-orbital), and $\tau = \pm1$ is the valley index. The wave functions of the valence band of the two valleys are now different. More precisely, they are related to each other by the time-reversal operator $(T\,|\psi^\tau\rangle = |\psi^\tau\rangle^* = |\psi^{-\tau}\rangle)$. The wave functions of the conduction bands are instead still identical.

Furthermore, in the vicinity of the K-valley of the valence band, there is a strong spin-orbit coupling due to the d-orbitals of the heavy-metal atoms. This translates in a huge spin-orbit splitting of the valence band of around 200 meV [24,25]. The spin-orbit splitting removes the spin degeneracy at each K and K' valley.

The two last properties together - inversion symmetry breaking and spin-orbit splitting - have as result that the K and K' valleys in the valence band are degenerate in energy but are populated with carriers with opposite spin.

This is understandable looking at the Hamiltonian of the system when the spin-orbit splitting $(\mathbf{L}\cdot\mathbf{S})$ is included. Using first-order $\mathbf{k}\cdot\mathbf{p}$ perturbation theory [8], in the basis given in Eq. (1.13), the Hamiltonian is:

$$H = at(\tau k_x \sigma_x + k_y \sigma_y) + \frac{\Delta}{2}\sigma_z + \lambda\tau\frac{\sigma_z - 1}{2}s_z,\tag{1.14}$$

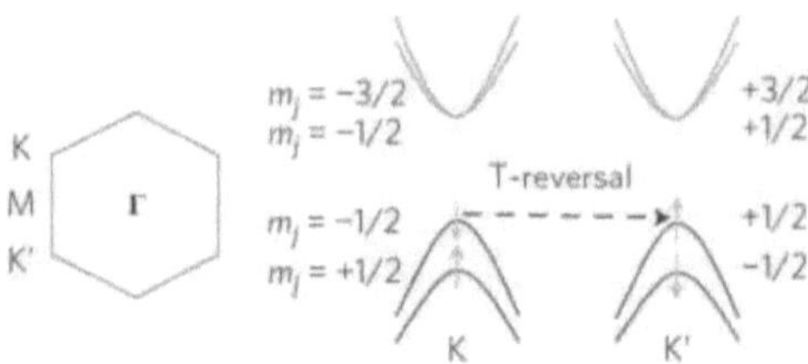

Figure 1.6.: Scheme of the Brillouin zone and the band structure in the vicinity of the K points of a TMD monolayer. The image was taken from [26].

where the last term is due to spin-orbit splitting, σ are the Pauli matrices in the basis of the wave functions, a is the lattice constant, Δ is the energy band gap, and t is the hopping integral. From Eq. (1.14), it is clear that the spin-orbit splitting is valley-dependent and the two highest energy points K and K' of the valence band are populated with carriers with opposite spin orientation. The scenario is represented schematically in Figure 1.6.

Due to the different spin population of the K and K' valley, there are valley-dependent selection rules for the optical polarization, i.e. absorption and emission [8,27]. More specifically, using circularly polarized light, it is possible to selectively excite carriers in only one valley (either K or K').

This can be seen calculating the matrix element of the interband transitions under circular polarized radiation. For left (+) or right (-) circularly polarized light ($P_{\pm} = P_x \pm iP_y$), using the Hamiltonian and the wave functions given above, one gets:

$$|P_{\pm}|^2 = \frac{m^2 a^2 t^2}{\hbar^2}\left(1 \pm \tau \frac{\Delta}{\sqrt{\Delta^2 + 4a^2 t^2 k^2}}\right)^2 \simeq \frac{m^2 a^2 t^2}{\hbar^2}(1 \pm \tau)^2, \qquad (1.15)$$

where the last approximation is valid for small k, i.e. close to minimum and the maximum of the conduction and valence band, respectively. From this equation, it is evident that the matrix element vanishes for one of the valleys if circularly polarized light is used (either left or right).

As a note, an essential property of the system for the valley-dependent optical polarization is the inversion symmetry breaking. This is obtained naturally in the monolayer limit. In a bilayer system, the inversion symmetry is restored and there is no valley dependency in the wave functions as well as in optical polarization.

The experimental demonstration of the valley-dependent optical absorption was carried out early on after the isolation of TMD monolayers [26, 28]. Since then, many other experiments were carried out using this property, e.g. the demonstration of valley-dependent Bloch-Siegert shift in WSe_2 monolayer [29] or valley Hall effect in MoS_2 monolayer [27]. Since this property holds very well in practical experiments, the idea of using the valley degree of freedom for quantum computation was also proposed. This is the basis for the so-called valleytronics.

1.2. Other two-dimensional semiconductors

As mentioned above, the library of two-dimensional materials is very big. Up to now, hundreds of different two-dimensional materials have been experimentally fabricated and many others have been predicted to be stable [30].

However, stability of materials in the two-dimensional form is a major issue for most of the compounds. Especially metal and organic compounds result to be very sensitive to air. Segregation, corrosion, and contamination by adsorbates are responsible for the degradation of the crystal and optical quality. In these cases, passivation of the materials is necessary. In some extreme cases, the isolation and passivation of 2D crystals is technically very demanding because it has to be done under ultra-high vacuum, as for instance for silicene [31].

Nevertheless, the set of two-dimensional materials is now very diverse. Most of the well-known opto-electronic, mechanical, and magnetic properties of three-dimensional solids have been reproduced in two-dimensional crystals. Figure 1.7 shows in a schematic view of the most studied two-dimensional materials.

Beyond graphene and TMD monolayers, one of the most used 2D-crystals in research is hexagonal boron nitride (hBN). Very commonly this material is used as a substrate or passivation layer for other van der Waals materials. In fact, it forms crystals with the same highly-compact honeycomb structure as graphite but it has a large band-gap energy [32]. For this reason, hBN is very resilient to interact with chemicals and gas molecules. At the same time, hBN is transparent in the visible and near-infrared: this fact makes hBN suitable to use it as passivation layer for spectroscopy. Furthermore, hBN is very flat, as all the other van der

Graphene family	Graphene	hBN 'white graphene'		BCN	Fluorographene	Graphene oxide
2D chalcogenides	MoS_2, WS_2, $MoSe_2$, WSe_2	Semiconducting dichalcogenides: $MoTe_2$, WTe_2, ZrS_2, $ZrSe_2$ and so on		Metallic dichalcogenides: $NbSe_2$, NbS_2, TaS_2, TiS_2, $NiSe_2$, and so on		
				Layered semiconductors: $GaSe$, $GaTe$, $InSe$, Bi_2Se_3, and so on		
2D oxides	Micas, BSCCO	MoO_3, WO_3	Perovskite-type: $LaNb_2O_7$, $(Ca,Sr)_2Nb_3O_{10}$, $Bi_4Ti_3O_{12}$, $Ca_2Ta_2TiO_{10}$ and so on		Hydroxides: $Ni(OH)_2$, $Eu(OH)_2$, and so on	
	Layered Cu oxides	TiO_2, MnO_2, V_2O_5, TaO_3, RuO_2, and so on			Others	

Figure 1.7.: Table with the most studied two-dimensional materials. The image was taken from [7].

Waals layered crystals, and this fact reduces the disorder potential induced by the roughness of other substrates, e.g. SiO_2. The use of hBN for encapsulation was firstly demonstrated in 2010 [33] and now it is widely used for most of the vdW crystals [34, 35]. But besides that, other interesting properties of hBN were pointed out, as for instance UV lasing or hyperbolic surface phonon-polariton with low losses [32, 36].

Other vdW semiconductors that attracted attention are black phosphorus [37], MOF films [38], IV-VI semiconductors [39], III-VI compounds as InSe and GaSe [40–43], and others. A general review on the recent development of two-dimensional materials can be found here [44].

During the course of this PhD, many experimental investigations were conducted on InSe and the results will be presented later on in the book. For this reason, a short introduction to the main properties of this material will be discussed in the next section.

1.2.1. III-VI semiconductors: InSe

InSe belongs to the III-VI semiconductors and can crystalize in different phases, as schematically shown in Figure 1.8. α-, β-, and γ-InSe have a layered crystalline structure [45]. Therefore, it is relatively easy to isolate thin-layer flakes down to the limit of monolayer. In this book, only γ-InSe was studied and only the properties of this phase will be discussed.

The unit cell is a four-layer structure of around 0.83 nm thickness [46, 47]. The

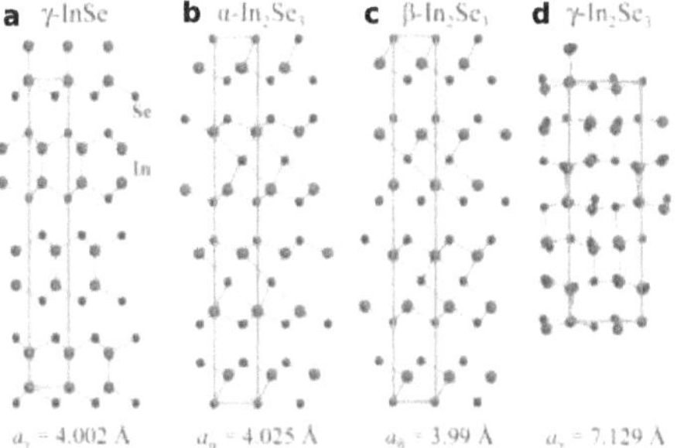

Figure 1.8.: Different phases of InSe. The image was taken from [45].

band structure depends strongly on the number of layers. Bulk InSe has a direct band gap of 1.25 eV around the Γ point and a monolayer of InSe has an indirect band gap at around 2.5 eV (Figure 1.9) [46,48]. The energy blueshift of the band gap with decreasing number of layers is a combination of reduction of dielectric screening and quantum confinement. The possibility to tune the band gap from near-infrared to visible is a promising feature for different applications, as for instance broadband detectors. The crossover from direct to indirect band gap takes place at around ten monolayer thickness.

A very peculiar aspect of the band structure is the mexican-hat shape of the electronic dispersion around the Γ point in the valence band if the material is less than ten layer thick and it is the reason of the indirect band gap. This feature leads to interesting effects, as tuneable magnetism and increase of the Seebeck coefficients [49,50].

Being part of the two-dimensional vdW semiconductors with direct band gap, thin-layer InSe got increasing attention by the scientific community in recent years. Most of the possible opto-electronic applications that use InSe as active material were already tested showing good performances, as for instance optical detectors [51], and bendable opto-electronics [52]. Recently, the possibility of using InSe for infrared applications was also pointed out [53].

Furthermore, InSe showed a high carrier mobility of the order of 10^3 cm^2/(Vs) due to the small electron effective mass of $m_e^* = 0.14\ m_e$ [48]. This fact induced researchers to investigate electronic devices, such as InSe-based field effect

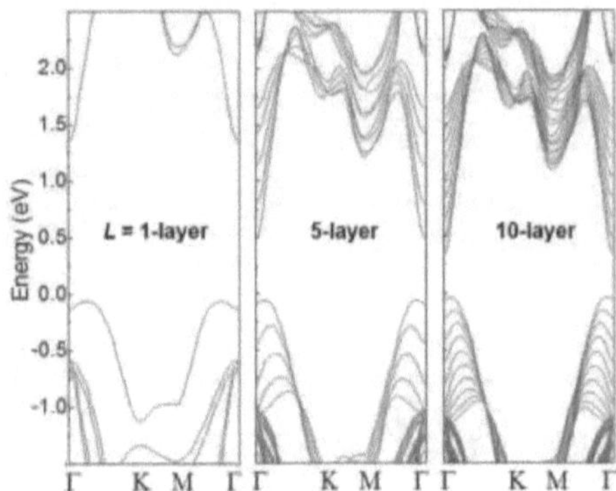

Figure 1.9.: Band structure of γ-InSe for different number of layers. There is a direct-to-indirect band-gap crossover with decreasing number of layers. The image was taken from [46].

transistors [54, 55]. A mobility of hundreds of $cm^2V^{-1}s^{-1}$ has been obtained [35].

Besides these promising properties, InSe degrades relatively quickly when it is exposed to air under ambient condition. Therefore, the passivation of this material is necessary both for practical applications and for scientific research. In Chapter 5, studies on InSe thin-flakes fully encapsulated in hBN will be discussed, with particular attention to the photoluminescence properties.

1.3. Van der Waals heterostructures

As mentioned above, the possibility to stack on top of each other different vdW materials without limitation on the lattice mismatch represents one of the most promising features of two-dimensional materials [7]. With this technique it is possible to create heterostructures with all the combinations of materials. Furthermore, combining atomically thin materials does not result in the sum of the properties of each of them, but there is a non-trivial interaction between the different layers leading to new exciting physics.

The fabrication of vertical heterostructures was firstly demonstrated in 2011 [56]. This heterostructure was a graphene-based device composed by two layers of graphene separated by a thin-layer of hBN. The fabrication of this device was

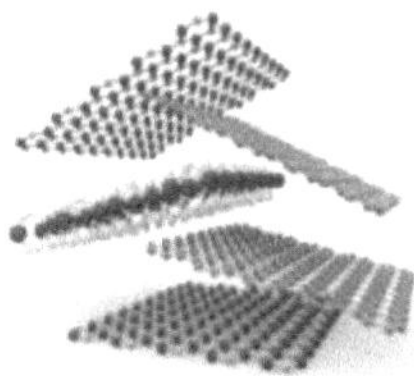 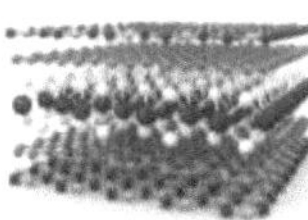

Figure 1.10.: Scheme of the concept of van der Waals heterostructures. Different layers are stacked on top of each other without constraint of lattice mismatch. The resulting heterostructure can combine in an extremely compact way the properties of each layer. The image is taken from [44].

done by means of a PMMA-assisted transfer of each layer.

In the meantime, many fabrication techniques have been developed. In terms of scalability, bottom-up approaches are favorable, as chemical vapour deposition [57]. However, up to now, the highest quality heterostructures are obtained by mechanical transfer. Furthermore, this last technique is completely versatile: there is no limitation regarding the use of different materials. Therefore, for research purposes, this approach is still favorable and is the one used for the fabrication of heterostructures presented in this **book**. A detailed description of this technique will presented in the next chapter (Section 2.1.2).

Famous examples of vdW heterostructures are the use of hBN as substrate and for passivation or the use of graphene on TMD as ohmic contact for electrical devices [33, 58]. But many other applications are of course possible and already demonstrated. The most natural application is the fabrication of pn-junctions. These devices can be composed by many different vdW materials and can be used as detectors, phototransistors, photovoltaic devices, or diodes [59].

Another very prominent and recent example of vdW stack is the homojunction of twisted bilayer graphene. Cao et al. showed that, via stacking two graphene monolayers on top of each other with a specific relative angle between the in-plane crystallographic axes ($\sim 1°$), it is possible to achieve superconductivity [60].

Besides the charm and the potential usefulness of the discovery, the experiment demonstrates that the bonds between single layers are not purely due to vdW forces but there is also a significant covalent contribution that can alter strongly the properties of the device.

The number of possible vdW heterostructures is clearly very large. The next section is dedicated to heterostructures of TMD monolayers. The special attention to this type of vdW heterostructure is due to the fact that WSe_2/$MoSe_2$ monolayer heterostructures were investigated during the last part of the PhD and the results are presented in chapter 6.

1.3.1. TMD monolayer heterostructures

TMD monolayers are one of the most investigated classes of vdW materials and, similarly, heterostructures made of these compounds are intensively studied. As mentioned above, the bonds between the two layers are mostly due to van der Waals interaction. Therefore, the coupling of the two layers is not very strong. However, there is also a covalent component that makes the heterostructure a non-trivial sum of the properties of each layer.

The first step to understand the physical properties of a heterostructure is looking at the band alignment. The band alignment of different TMD semiconductors is shown in Figure 1.11. Differently from other semiconductor heterostructures, there is no band bending because of the missing chemical bonding at the interface and the low space charge effect. Therefore, the band alignment is relatively easy to obtain with decent accuracy just using Anderson's rule [61].

In this **book**, heterostructures of $MoSe_2$ and WSe_2 monolayers are studied, therefore the discussion will be focused on this type of heterostructure. However, the considerations are also significant for heterostructures made out of other TMD semiconductors.

$MoSe_2$/WSe_2 heterostructures have staggered band alignment (type II), as can be seen in Figure 1.11. The band distribution is such that, after optical excitation, there is a charge transfer between the two layers towards the lowest energetic state. The $MoSe_2$ layer results negatively charged and WSe_2 positively charged. This charge transfer creates a net dipole moment in the heterostructure.

The charge transfer is extremely fast: it occurs on a time scale of hundreds of femtoseconds [63]. After the fast charge transfer, the holes and the electrons in the different layers interact strongly leading to the formation of electron-hole bound pairs. These pairs are called interlayer exciton. The formation of interlayer

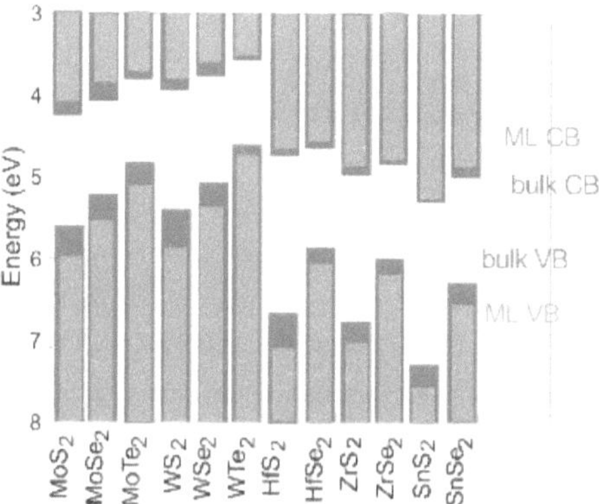

Figure 1.11.: Band alignment of TMD materials obtained using Anderson's rule. The image is taken from [62].

excitons is also very fast. Merkl et al. measured the interlayer exciton formation and dynamics detecting the intraexcitonic transitions of the interlayer exciton and getting an exciton formation time of few picoseconds [64].

Due to this fast charge transfer, the interlayer excitons have a permanent out-of-plane dipole moment. This results in a repulsive interaction between interlayer excitons. Due to this repulsive interaction, the formation of exciton complexes like interlayer trions or biexcitons is not probable, even at high exciton densities. Together with the high interlayer exciton binding energy (~ 100 meV), this aspect has led to the observation of exciton condensation at temperatures as high as 100 K [65].

Furthermore, a feature that has to be taken into account when vdW heterostructures are studied is the relative orientation of the crystal lattices. MoSe$_2$ and WSe$_2$ monolayers have the same crystal structure but two slightly different lattice constants. Due to the lattice mismatch of the two compounds, a moiré potential is formed in the heterostructure even when the two lattices are perfectly aligned. The moiré potential can lead to the formation of localized interlayer excitons with a very long lifetime [66]. Moreover, the moiré potential can be controlled changing the relative angle of the two lattices, i.e. the twist-angle. This is a unique feature that has already shown intriguing physics. A more detailed description of the exciton physics in TMD heterostructures will be given in Chapter 6.

As becomes clear from this short discussion, a lot of interesting physics has been already discovered in TMD heterostructures. However, a strong driving force for the research on this topic is the promising results on opto-electronic applications. Photo-transistors, photo-voltaic devices, and emitting photo-diodes have been already demonstrated with excellent performances [67–69]. Furthermore, these devices can be used for specific applications taking advantage of the mechanical properties of these extremely thin heterostructures, as for instance bendable electronics or semi-transparent devices.

2. Sample fabrication and experimental techniques

In this chapter, the experimental techniques used in the experiments presented in this **book** are discussed. The chapter is divided into three sections which treat the following topics:

- Sample fabrication. First, a brief discussion on the state-of-the-art and outlook of growth and fabrication techniques of van der Waals materials is discussed. Thereafter, a detailed description of the exfoliation of mono- and few-layer flakes is given, followed by the description of the dry-transfer technique used for the fabrication of heterostructures.

- Spectroscopic techniques used for the most relevant experiments. The concepts and the technical details of the experimental setups are described. The techniques are photoluminescence (PL), time-resolved PL, and pump-probe spectroscopy.

- The infrared Free-Electron Laser (FEL). A description of the working principle of the FEL is presented. Afterwards, the specific details of the infrared free-electron laser at HZDR (FELBE) are discussed.

2.1. Sample fabrication

The fabrication of good quality two-dimensional van der Waals materials is one of the main technological challenges before the implementation of real-life applications. The oldest example of two-dimensional material is graphene which was isolated for the first time in 2004 by mechanical exfoliation. Even though many possible applications have been pointed out and some of them are in use, the fabrication of high-quality graphene for mass production is still a topic of intense research.

The same technological issues are faced by TMD monolayers which is a younger family of van der Waals materials. Up to now, the fabrication technique that provides the highest quality of samples is the mechanical exfoliation of bulk crystals. This technique is quite efficient for the production of few high-quality samples, i.e. for scientific research, but clearly not scalable for mass production.

Another considerable drawback of mechanical exfoliation is the small in-plane size of the monolayer flakes: the lateral dimension of a flake is typically limited to tens of micrometers. This point is particularly critical for the study of infrared response of TMD monolayers since the spot diameter in the far-field is limited by the wavelength of the radiation.

For these reasons, a very active field of research is the development and optimization of the growth of van der Waals mono- and thin layers. Interesting alternatives to mechanical exfoliation are liquid-phase or chemical exfoliation [70, 71]. These techniques provide good quality samples and are scalable for mass production. In addition, a lot of progress has been achieved in the development of these techniques leading also to effective functionalization and applications [72, 73].

A more favorable approach for scalable production is the use of bottom-up techniques. The most investigated technique for the growth of TMD monolayers is chemical vapor deposition (CVD). The first production of CVD-grown TMD monolayer was demonstrated in 2012 [74]. Since then, many progresses have been done varying the growth parameters as temperature, pressure, precursors and substrates [75]. Furthermore, several different CVD approaches have been tried out as for instance atomic layer deposition [76], metal-organic CVD [77], molecular beam epitaxy [78], physical vapor deposition [79], and pulsed laser deposition [80], reaching very good crystal and optical quality, high carrier mobility, and large-area monolayer samples with good uniformity.

Nevertheless, most of the discoveries on TMD monolayers and on other atomically thin materials have been made using mechanical exfoliation as fabrication technique. Mechanical exfoliation is the technique that has been used for the experiment presented in this book. Therefore, a step-by-step description of the fabrication process is presented in the next section.

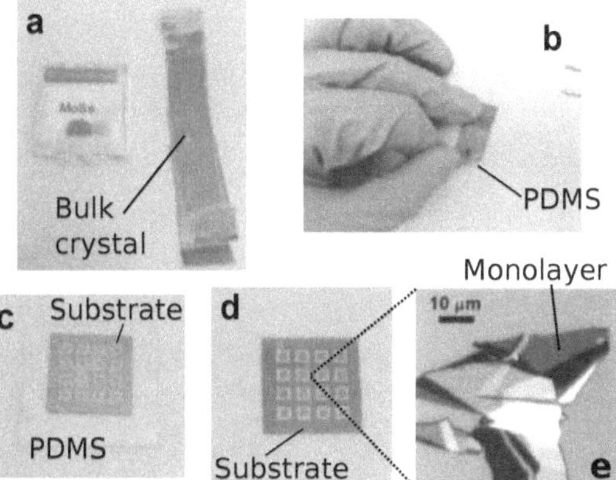

Figure 2.1.: (a) A small piece of a bulk crystal is put between two adhesive stripes. (b) After few iterations, the adhesive tape with some material is put on top of a PDMS clean foil and quickly peeled off. The monolayer flake is formed. (c) The PDMS foil is transferred on a desired substrate. (d) After the removal of the PDMS foil, some material is transferred on the substrate. The biggest flakes are barely visible by eye. (e) A monolayer flake is visible with an optical microscope. The substrate is $SiO_2(280 \text{ nm})/Si$.

2.1.1. Mechanical exfoliation

All the tools necessary for the preparation of monolayer flakes by means of mechanical exfoliation are a high-quality bulk crystal, adhesive tape, a transparent adhesive plastic foil (Polydimethylsiloxane (PDMS)), and a substrate.

For the fabrication of TMD samples, the crystals were bought from the company *2D Semiconductors*. The crystals are grown by chemical vapor transport and the company assures that they are pure at 99.9999%. InSe crystals were provided by Amalia Patanè's group from University of Nottingham and they were grown by Bridgman technique. The hBN bulk crystals were provided by Taniguchi and Watanabe from the Advanced Materials Laboratory National Institute for Materials Science in Tsukuba and they were grown under high pressure and high temperature [81].

The adhesive tape is made of a polyvinyl chloride (RPVC) film prepared in

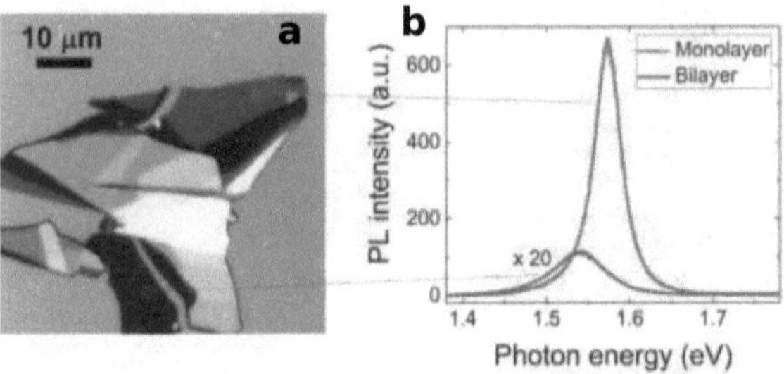

Figure 2.2.: (a) Optical image of MoSe$_2$ flakes exfoliated on SiO$_2$(280nm)/Si. (b) PL spectra of MoSe$_2$ monolayer and bilayer.

a clean room environment. The name of the adhesive tape is *SWT 20+R* and was bought by the company *Nitto*. Other types of adhesive tapes were tested with similar results in terms of fabrication yield and optical quality of the final samples. The most important characteristics of the the adhesive tapes are the cleanness and the stiffness. The PDMS foils were bought from *Gelpak*. Again, the cleanness of the PDMS foils is absolutely necessary.

The first step is putting a small piece of bulk crystal between two stripes of adhesive tape and peeling off the two stripes (Figure 2.1(a)). After few iterations, when the amount of material on the tape is significantly less but still visible by eye, a stripe of the adhesive tape is put on top of a small piece of PDMS foil and some pressure is applied by hand (Figure 2.1(b)). At this point, putting the foil with the adhesive tape on a heater at around 50°facilitates the formation of monolayer flakes. Another equally efficient possibility is to wait for at least half an hour before proceeding with the next step.

Therefore, after waiting some time, the adhesive tape is peeled off quickly from the PDMS foil (Figure 2.1(b)). In this step, the monolayer crystals are isolated. In order to find the monolayer flakes on the PDMS foil, an optical microscope in transmission mode is used. The optical contrast of the monolayer flakes on PDMS is rather poor, therefore it is necessary to increase the optical contrast of the microscope saturating the contrast scale.

The monolayer flakes can be identified just by looking at the optical contrast in a microscope. However, to further confirm the thickness of the flakes, photo-luminescence (PL) and Raman spectroscopy are used. In particular, PL is the ideal tool for TMD monolayers since these materials have direct band gap only in the monolayer limit [11]. The PL emission is strongly enhanced if the flake is monolayer.

At this point, the monolayers can be transferred onto different substrates or on other van der Waals materials for the fabrication of heterostructures (Figures 2.1(c) and (d)). However, $SiO_2(280nm)/Si$ was often used as substrate in order to maximize the optical visibility of the monolayer flakes [82, 83].

Figure 2.1e shows a typical optical image of vdW materials exfoliated on $SiO_2(280$ nm)$/Si$ substrate. In this example, $MoSe_2$ is used. It is already possible to recognize thinner flakes. PL spectra measured in a monolayer and bilayer region are shown in Figure 2.2(b). The PL spectra of bilayer and monolayer $MoSe_2$ are not only different in intensity (roughly two orders of magnitude higher PL yield for monolayer flakes), but there is also an energy shift (~ 30 meV). Other TMD materials have a similar behavior [84]. This measurement was performed using the micro-PL setup described later (Section 2.2.1).

Another useful and non-invasive tool to double-check the thickness of the exfoliated flakes is Raman spectroscopy. Raman spectra of WSe_2 flakes with different thicknesses are shown in Figure 2.3. The Raman peak at 309 cm^{-1} is the B_{2g} interlayer phonon mode and, therefore, is observable only if the flake has more than one layer. The absence of this peak in the green curve in Figure 2.3 is a very clear fingerprint that the flake is monolayer.

2.1.2. Fabrication of van der Waals heterostructures

The fabrication of van der Waals heterostructures is more complicated but it is again based on very simple technology. The development of deterministic transfer of exfoliated two-dimensional materials was firstly demonstrated for graphite and other nanostructures in 2006 [85] and afterwards applied to all the other two-dimensional compounds [86]. The fabrication of the van der Waals heterostructures presented in this book was carried out by the same PDMS-based dry-transfer method. A short description and an example are reported below.

The vdW materials are firstly exfoliated on PDMS foils as described above. It is crucial that the foil is transparent, but other plastic materials can be used. After the identification of the flakes on PDMS, the heterostructures are ready to

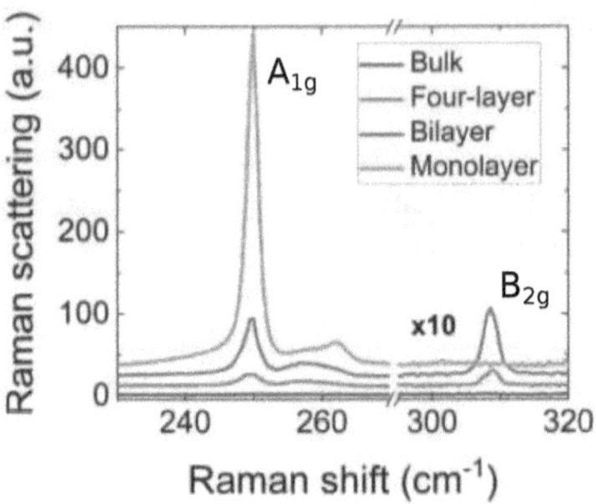

Figure 2.3.: (a) Layer-dependent Raman spectra of WSe$_2$ flakes. The spectra are shifted vertically for clarity. In Figure b, the absence of the interlayer phonon in the monolayer spectrum is a fingerprint of the single-layer nature of the flake.

be assembled. The idea is to stamp the flake on PDMS on top of the other flake on the substrate looking through the PDMS foil.

The stamping procedure is carried out using an optical microscope coupled with two motorized stages that are free to move in three dimensions. The PDMS foil is positioned on a transparent area of the upper stage, so that it is possible to look through the PDMS and look simultaneously both at the flake on PDMS and at the flake on the substrate. When the two flakes are on top of each other, the PDMS foil is delicately stamped on the target.

An example of heterostructure fabrication is shown in Figure 2.4. Different phases of the stamping process are captured with the optical microscope. The heterostructure is hBN/InSe/hBN.

The most delicate moment is when the two flakes are in contact and the PDMS foil has to be removed. During this step, the flake can break or can stay attached to the PDMS. To facilitate the adhesion of the two flakes, the temperature of the stage is often increased up to 50 C°when the two flakes are in contact. Another crucial detail is to remove the PDMS foil extremely slowly having a look constantly at the sample with the optical microscope, as in Figure 2.4(c).

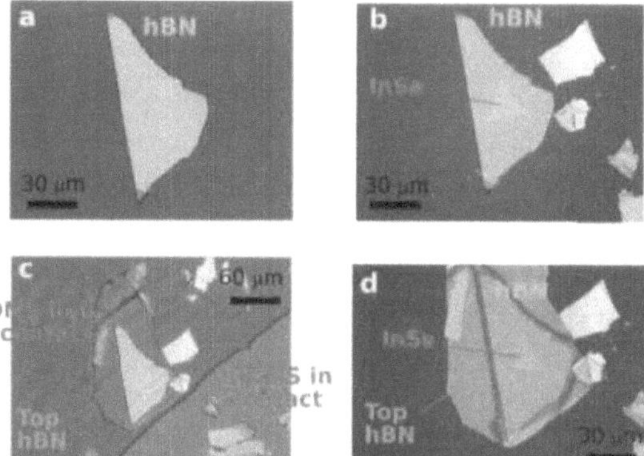

Figure 2.4.: (a) Bottom hBN layer on SiO$_2$/Si substrate. (b) InSe flake already placed on top of a hBN flake. (c) Microscope picture at the moment of the stacking of the top hBN flake. The PDMS is only partially in contact with the substrate and a clear difference in the color is visible because of optical interference. (d) Final hBN/InSe/hBN heterostructure.

An useful feature of the micro-manipulator is the possibility to rotate the substrate while the PDMS stamp does not move. In such a way, it is possible to control with very good accuracy the relative angle between the two flakes. In some cases, as for instance for TMD monolayer heterostructures, the control of the stacking angle, i.e. the relative orientation of the crystallographic axes or twist-angle, is a fundamental parameter that has to be kept under control.

For instance, the stacking angle is the parameter that drives and determines the moiré superlattice formation [87]. The associated moiré potential has several implications for different systems, as the localization of interlayer excitons in MoSe$_2$/WSe$_2$ heterostructures [66].

Of course, the stamping can be done also on some special substrate as for instance grids, plasmonic or photonics devices, or metal contacts.

2.2. Experimental techniques for spectroscopy

The scope of this book is the investigation of the optical and infrared properties of van der Waals semiconductors. In order to do that, different techniques of spectroscopy have been used. The most relevant techniques for the experiments presented later are photoluminescence and pump-probe spectroscopy. The general principle of these techniques as well as the technical details of the setups are reported in this section.

2.2.1. Photoluminescence

Photoluminescence is a physical phenomenon of spontaneous light emission from a material induced by absorption of light. Photoluminescence originates from a pure quantum mechanical mechanism. Microscopically, a particle in an excited state can relax to a lower energy state via emission of a photon. This spontaneous radiative emission is called luminescence. When the excitation of the particle is induced by absorption of light, the phenomenon is called photoluminescence.

In solid state physics, photoluminescence is often used to study the electronic transitions in direct band-gap semiconductors. Also in this book, photoluminescence was used as one of the main spectroscopic tools to investigate the optical properties of direct band-gap two-dimensional semiconductors.

A simple analytical expression for PL can be given with the Elliott formula for photoluminescence:

$$I_{PL}(\omega) \sim \mathrm{Im}\left(\sum_n \frac{|\psi_n(r=0)|^2 S_n}{E_n - \hbar\omega - i\delta_n} \right), \tag{2.1}$$

where the sum is over the excitonic states, ψ_n are the excitonic wavefunctions, and S_n is the spontaneous emission source [88]. The spontaneous emission term contains an exciton and a free-electron-hole contribution, i.e. $S_n = S_n^{ex} + S_n^{e-h}$. Both terms add a radiative contribution at the same photon energy, therefore they are not spectrally distinguishable only looking at the energy position.

Equation (2.1) can be derived starting from the semiconductor luminescence equations and the full derivation can be found in textbooks [89]. However, a simple and instructive way to derive the PL Elliott formula is from the Elliott

formula for absorption (see appendix, Eq. (A.5)). In fact, the two formulas have
the same form. One can compactly write:

$$I_{PL}(\omega) \sim \alpha_X S_n. \tag{2.2}$$

This equation connects directly absorption and incoherent emission in a semiconductor, in analogy to the detailed balance of the Einstein coefficients for transitions. Emission and absorption are connected via the thermal population. Absorption of radiation and PL emission are strictly connected and both carry direct
information on the medium in a complementary way.

In the next two sections, the operation and the technical details of the different
photoluminescence setups are described.

Micro-photoluminescence

Electromagnetic radiation can be focused at best down to the diffraction limit,
i.e. $\Delta x \geq \frac{\lambda}{2}$, where λ is the wavelength of the radiation and Δx is the spot
diameter. By using spatially coherent radiation, it is however feasible to approach
this limit. Therefore, with standard far-field techniques, it is possible to collect PL
from an area as small as the wavelength of the radiation used for excitation (if the
wavelength of the excitation is longer than the emitted wavelength, the maximum
spatial resolution is limited by the emitted wavelength). Using visible light as
excitation, it is possible to get a spatial resolution on a scale of micron or even
less. In this cases, the spectroscopic technique is called micro-PL. This technique
is particularly useful for the samples presented in this **book**: the samples have a
lateral size on micrometer scale.

A standard and useful technique is to get a PL topography map of a sample
with spatial resolution in the micrometer range. This technique is applied in solid
state physics but also in many other fields of research as for instance biological
samples.

The setup of micro-photoluminescence used in the experiments presented later
is a commercial system bought from *HORIBA*. The excitation source is a cw
frequency-doubled Nd:YAG laser at $\lambda = 532$ nm wavelength. The spot diameter
on the sample is around 1.5 μm. The laser is focused on the sample with a 100x
magnification objective with a numerical aperture of 0.8. The sample is mounted

Figure 2.5.: Picture of the micro-PL setup arranged for low-temperature measurements.

on a motorized stage in such a way that is possible to perform PL mapping. The PL emission is collected by the same objective in a back-scattering geometry and it passes through the same notch filter that was used as injection filter. After filtering of the stray light by a long-pass filter, the PL is dispersed in a 300 grooves per millimeter grating onto a nitrogen cooled Si-CCD deep-depletion camera.

The same instrument was used for Raman spectroscopy in order to characterize the samples as discussed in Section 2.1. In this case, a grating with 1800 grooves per millimeter was used in order to get a better spectral resolution.

For low-temperature PL spectroscopy, the same HORIBA setup was used coupled with a cold finger cryostat (Figure 2.5). In this case, the motorized stage was removed from the commercial setup and the cryostat was mounted on a micro-manipulator holder.

2.2.2. Time-resolved micro-photoluminescence

In order to study the dynamics of the PL emission, a pulsed laser and a fast detector have to be used. The idea is that a short pulse of light excites the system, for instance carriers in the conduction band of a semiconductor. The PL emission, that is proportional to the excited population, will be vanishing in time. Long enough after the arrival of the excitation pulse, no PL emission will be detected. This technique gives direct information on the dynamics of the photo-excited population and, therefore, the subsequent radiative and non-radiative decay.

This technique is not only useful for the understanding of the PL dynamics but can be crucial, for instance, in order to distinguish two spectroscopic features that were not spectrally resolvable or to assign a physical meaning to some emission bands.

The time resolution is usually limited by the detector system. For the time-resolved PL measurements presented in this **book**, two different detection techniques were used depending on the range of the time constants of the PL decay. For decay time constants between 2 ps and 2 ns a streak camera was used. For decay constants up to 200 ns a single photon counting technique was used.

Even if the detection technique is different, the arrangement of the two setups was the same both in terms of optics and geometry. A mode-locked Titanium:Sapphire oscillator with pulse length of 3 ps was used as excitation source with wavelength between $\lambda = 750$ nm and $\lambda = 840$ nm. For some experiments, second harmonic generation (SHG) of the laser was used as excitation source. The SHG was obtained by focusing the laser onto a LBO crystal mounted on a thermally controlled system. The repetition rate of the laser is 78 MHz.

The excitation beam is focused on the sample and the PL was collected and collimated through an achromatic Mitutoyo 100x magnification objective. The PL was afterwards brought to the detection systems. The two detection systems are described separately in the next sessions.

Streak camera

In order to record time-resolved PL spectra, an ultrafast detector system is needed. A streak camera is one of the fastest optical detector available overall. There are commercially available streak cameras with time resolution up to 100 fs.

The operating principle of a streak camera is shown schematically in Figure

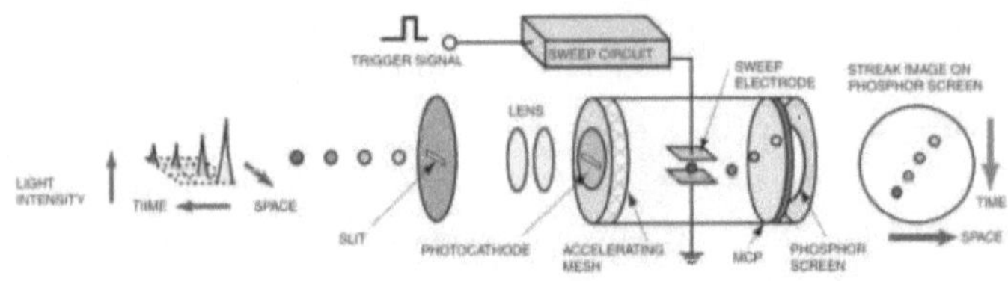

Figure 2.6.: Operating principle of a streak tube [90].

Grating (grooves/mm)	Spectral window (nm)	Spectral resolution (nm)	Temporal resolution (ps)
100	160	2.0	5
300	50	0.6	15
1800	12	0.1	40

Table 2.1.: Parameters of different gratings.

2.6. The PL radiation is dispersed in a spectrometer and afterwards focused into the streak tube. The photons are converted into electrons with a photocathode. The accelerated electrons are deflected with a fast oscillating signal in the vertical direction and recorded on a CCD camera. The oscillating signal is triggered by the excitation laser so that there is a stable synchronization between the PL photons and the detection time.

At the same time, the PL radiation is dispersed on the horizontal axis so that it is possible to get the spectral information and the time information at the same time.

The streak camera system used in the experiments described in this book is the Hamamatsu C5680 system that is optimized for broadband visible and near-infrared detection at 78 MHz, i.e. the repetition rate of the excitation laser. The minimum time resolution is around 4 ps. The streak tube is coupled with a spectrometer comprising three different gratings with different characteristics, as listed in Table 2.1.

The CCD camera and the optics are optimized for a broad spectral response from 300 nm to 1600 nm. Figure 2.7 shows the quantum efficiency of the streak camera measured in the lab.

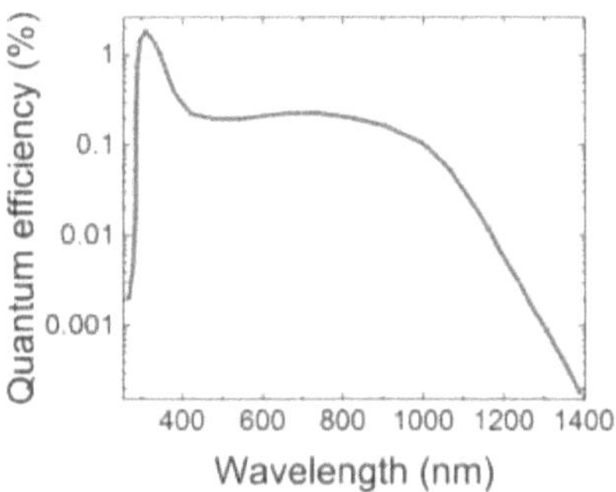

Figure 2.7.: Quantum efficiency of the streak camera.

Time-correlated single-photon counting

Time-correlated single-photon counting is a technique commonly used to measure time-resolved PL. The concept is to measure the correlation between a pump pulse that induces the PL from a sample and a reference pulse from the same laser used as a clock. At every pulse arrival at the reference, one photon count of PL is recorded. From this condition, it is clear that a single-photon fast detector is needed for the detection of the PL radiation. In contrast, for the reference there is no need for single-photon detectors.

A necessary condition is that the average number of PL photons per count at the detector is less than one. Otherwise, if more than one photon per pulse arrives at the detector, the system will record only the photon that arrived earlier, leading to a distortion of the PL decay and an underestimation of the time constant (pile-up effect).

A sketch of the setup used in the experiments is depicted in Figure 2.8. A single-photon avalanche diode was used as detector. The time resolution is typically limited by the detector itself. Here, the minimum time resolution is around 50 ps. The quantum efficiency of the detector is 4%. The fast detector is coupled with a monochromator where the PL is collimated into. The monochromator is a blazed grating and a slit. Different gratings are available with different spectral resolution. The spectral resolution of the monochromator is determined by the grating and the slit width. With this system it is possible to get good time

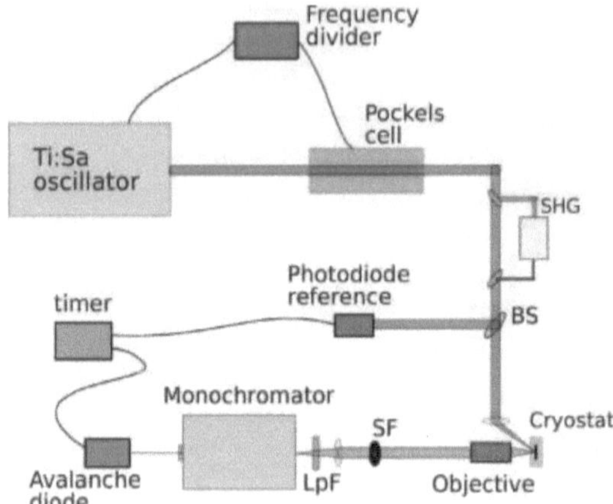

Figure 2.8.: Sketch of the time-correlated single-photon counting setup. LpF is a long pass filter. SP is a spatial filter. BS is a beam splitter.

resolution, high sensitivity, and excellent spectral resolution.

The minimum PL lifetime that is possible to measure is limited by the time resolution. On the other side, the PL lifetime should not be longer than the distance between two pulses, i.e. the inverse of the repetition rate of the laser. The repetition rate of the excitation laser in the setup is 78 MHz that corresponds to a period of 13 ns.

In order to extend the maximum time range, the repetition rate of the laser can be reduced with a pulse picker. The pulse picker is a Pockels cell. A fast electrical pulse is sent in the Pockels crystal when the beam is traveling into it. The electrical pulse is shorter than the distance between two laser pulses. The electrical pulse induces a birifrangence in the crystal and changes the polarization of the pulses. A polarized beam splitter is then used to select the pulses with a certain polarization.

The repetition rate is therefore controlled by the frequency of the electrical pulses. Of course, this frequency has to be commensurate with the repetition rate of the laser. The electrical pulses comes from a fast photodiode that measures the original laser pulses. A frequency divider is then used and the electrical pulses with the desired repetition rate are sent into Pockels cell.

As a final comment, the instrument response function (IRF) of the setup is not Gaussian. The most accurate way for the estimation of the decay constants is to measure the pulse of the excitation beam, i.e. the IRF, and to convolute it numerically with the chosen PL decay function.

2.2.3. Pump-probe spectroscopy

The essential component of a time-resolved photoluminescence setup is a fast detection system. In contrast, pump-probe time-resolved spectroscopy uses a different concept and there is no need for fast detectors. The idea is to use two beams, one is used to perturb the system (pump) and the other to measure the changes induces by the pump (probe). Using short pulses and changing the relative arrival time on the sample of the two beams, it is possible to record the changes induced by the pump beam as a function of time, i.e. the evolution in time of the system after the arrival of the perturbation.

In order to change the arrival time on the sample, the pump (or the probe) beam passes through a mechanical delay stage. The delay stage changes the beam path length simply moving two mirrors without changing the rest of the beam path. The delay stage is usually motorized, therefore the position of the mirrors can be controlled easily with micrometric precision. A precision of 1 μm of the delay stage corresponds to a delay of 3 fs of arrival time of the pulse on the sample.

In pump-probe spectroscopy, the time resolution is usually limited by the pulse length of the radiation. With this technique, attosecond time resolution has been achieved using x-ray radiation [91].

For the scope of this **book**, pump-probe spectroscopy has been used in different spectral regions with different sources and setups. As near-infrared pulsed laser a Titanium:Sapphire oscillator was used. The free-electron laser was used as pulsed infrared radiation.

When two laser sources at two different frequencies are used, the experimental technique is usually referred as two-color pump-probe spectroscopy. For this **book** a infrared-pump (FEL) near-infrared (NIR) probe experiment was performed and it will be presented later. A tricky issue in this case is the time synchro-nization of the laser sources. The pulses have to arrive at a precise relative time distance on the sample and this time relation has to be kept stable over time.

First of all, the lasers have to be synchronized. A FEL 13 MHz signal was used as master clock for the near-infrared laser (78 MHz, i.e. one FEL pulse every six NIR pulses). A synchronization unit adjusts the cavity length of the oscillator controlling the position of a mirror on piezo-electric stage to finely control the repetition rate and get a perfect synchronization with a very small jitter (around 1 ps).

Once that the sources are synchronized, in order to make the pump and probe pulses arrive on the sample at the same time, an electronic delay is used for rough adjustment and a mechanical delay stage that changes the beam path is used for the fine control. For this procedure, fast detectors and a fast oscilloscope are needed. For infrared radiation, a GaAs-based plasmonic detector was used. A silicon photodiode with large bandwidth (25 GHz) was used for near-infrared radiation.

As mentioned above, the detector used for measuring the probe beam does not have to be ultrafast. However, since the pump beam is usually modulated with a mechanical chopper at around 200 Hz, the bandwidth has to be larger than this value. For detection of NIR radiation, a silicon photodiode was used coupled with an amplifier. For detection of infrared radiation, either an MCT detector or a bolometer was used, depending on the wavelength region to detect.

Generally, the instrument response function in pump-probe measurements is given by the convolution of pump and probe pulses. These can be reasonably assumed to be Gaussian. Therefore, the convolution is also a Gaussian function whose variance is the pythagorean sum of the variances of the two pulses. The variance of the IRF is therefore given by $\sigma_{IRF} = \sqrt{\sigma_{pump}^2 + \sigma_{probe}^2}$.

In this case the decay function for fitting can be written analytically with the convolution of the IRF with the chosen decay function $G(t)$ i.e.:

$$I(t) = IRF(t) * G(t) \propto \int d\tau e^{-\frac{(\tau - t_0)^2}{2\sigma^2}} G(\tau - t), \tag{2.3}$$

where t_0 is the overlap time of the two pulses. The example of exponential decay is reported in Appendix A.2.

The time resolution is usually assumed to be full width at half maximum of the IRF, i.e. $\Delta t = 2\sqrt{2\ln(2)}\sigma_{IRF} \simeq 2.4 \cdot \sigma_{IRF}$.

2.3. Free-electron laser

A crucial part of the doctoral work was carried out at the infrared free-electron laser FELBE. The most significant experiments for this book conducted with the FEL will be presented later. Other experiments have been performed on different systems and setups and, even if they are not described in this book, they are an important part of my personal educational background. Since the use of a FEL is so a fundamental and distinctive topic of the PhD, this section is dedicated to the description of the general working principle of a FEL and of the specific characteristics of FELBE.

2.3.1. Overview

A free-electron laser is a laser based on acceleration of electrons in a magnetic field. This technology was developed for the first time in the late 70s at Stanford University and is in use still today [92]. Nowadays, there are many free-electron lasers operating from the THz to the hard x-rays spread all over the world.

The greatest advantages of this laser system are the extremely high brilliance and large wavelength tunability. The drawbacks of this kind of source are the size and the cost for operation that are insurmountable limits for the use of FEL in commercial applications and for widespread research.

The operating principle of a FEL is based on technology and on understanding of physics that was developed along the course of last century.

The electrons are generated usually with thermoionic emission and, afterwards, collimated and linearly accelerated to almost the speed of light. The electron beam is then guided into an undulator. An undulator is a double array of magnets with opposite polarity arranged in a periodic structure. The electron beam is therefore deflected in a sinusoidal trajectory by Lorentz forces, as shown schematically in Figure 2.9.

The deflection of electrons produces emission of photons (monochromatic synchrotron radiation). The synchrotron radiation is collimated in an optical cavity. In the case of x-ray FEL, no optical cavities are used but the lasing is obtained with the use of self-seeded techniques and very long linear acceleration. The photons are emitted almost parallel to the propagation direction of the electron

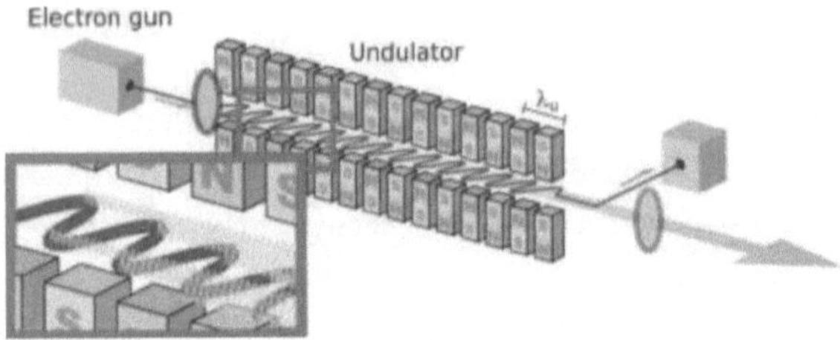

Figure 2.9.: Sketch of the main components of a free-electron laser adjusted from [93].

beam (relativistic effect) and they start to interact with the propagating electron beam. More precisely, the electric field associated with a photon accelerates or decelerates electrons according to the relative phase of the electron and the photons. This leads to the formation of bunches of electrons in the optical cavity (micro-bunching). Since the electrons are organized in micro-bunches, they start to emit coherent light that constructively interferes. This increase of coherent emitted radiation helps further the micro-bunching. This self-amplified process induces lasing.

After the micro-bunching, the light emission from the accelerated electrons is spatially and temporally coherent.

In a classical definition, a laser source has to be made of an active medium, a pumping system and a cavity. In a FEL, the electrons and the undulator can be considered to be the active medium and the electron accelerator is the pumping system.

The emission wavelength of a free-electron laser is given by the simple formula:

$$\lambda_{FEL} = \frac{\lambda_{und}}{2\gamma^2}\left(1 + \frac{K^2}{2}\right) \tag{2.4}$$

where λ_{und} is the period of the array of magnets, $\gamma = \left(1 - \frac{v^2}{c^2}\right)^{-\frac{1}{2}} = \frac{E}{mc^2} + 1$ is the relativistic factor that includes the electron beam velocity or energy, and $K = \frac{eB\lambda_{und}}{2\pi m_e c}$ [94].

From Equation (2.4), it is clear that it is possible to tune the emission wavelength of the FEL by tuning the periodicity of the undulator, the electron beam energy, and the magnetic field of the undulator. The tunability of the emission wavelength is one of the main advantages of a FEL. Taking advantage of this property, it is possible to obtain intense laser sources at wavelengths that are difficult to reach with other types of lasers, e.g. solid-state or gas lasers.

The only limitation in wavelength is at the high frequency side. The first technological problem is the lack of optics available in the x-ray regime. In fact, x-rays have energies higher than the electronic interband transitions in solids (no dielectric mirror) or the plasma frequency in metals. For this reason, it is not possible to build an optical cavity. To overcome this problem, a sufficiently long FEL has to be fabricated in order to obtain sufficient amplification over one pass of the electron into the undulator. This process is called self-amplified spontaneous emission.

Another advantage of the free-electron lasers is the high brilliance of the radiation due to the high coherence of the electrons. Simply speaking, the emitted radiation is proportional to the square of the charge e^2. If the radiation is incoherent, the emitted power would be $P \sim Ne^2$. If the radiation is coherent, it would be $P \sim (Ne)^2$. Furthermore, the emission angle of the radiation is very small due to relativistic effects of the electron beam.

The free-electron laser at ELBE (FELBE) covers the range of mid- and far-infrared. A general introduction and some physical key features of FELBE will be discussed in the next section.

2.3.2. FELBE

The free-electron laser at HZDR (FELBE) had its first operation in 2005 and since then a large number of experiments have been performed [95]. The FEL operates in the infrared range of wavelengths, more specifically between $5-250$ μm ($60 - 1.2$ THz). This broad range of wavelengths is covered using two undulators with different periodicity (37 mm and 100 mm for short and long wavelength, respectively).

In this range of frequencies, many physical resonances are active in the solid state of matter, e.g. lattice vibrations, magnons, intersubband transitions. Considering also that THz technology is a growing field for applications, an infrared FEL is an important facility both for scientific basic research and for the development of infrared technologies.

From the technological point of view, an infrared FEL is particularly relevant because of the so called THz-gap. Up to now, microwave technology cannot reach easily THz frequencies and, at the same time, optical and infrared technology cannot reach easily frequencies as low as few THz. The result is that there is a lack of THz sources of coherent light in the region between 0.1 and 10 THz. In contrast, an infrared FEL provides a tunable and intense lasing source that covers the entire range of infrared.

FELBE is based on a superconducting linear electron accelerator at around 40 MeV. The electron beam has an average current of 1 mA. A very special feature of FELBE is the continuous-wave operation made possible by the superconducting system.

The FEL emits infrared pulses with a pulse length of $1-10$ ps depending on the emission wavelength. The pulse repetition rate (PRR) is fixed to $PRR = 13$ MHz. This value fixes also the length of the optical cavity that is $L = \frac{c}{2 \cdot PRR} = 11.5$ m.

The high repetition rate and the high power in the THz gap are the characteristics that makes FELBE a unique laser source in comparison with the available table-top lasers.

3. Adsorbed gas molecules and laser irradiation in MoSe$_2$ monolayer

In this chapter, an investigation of the influence of adsorbed gas molecules and laser irradiation on the optical properties of MoSe$_2$ monolayer is presented. After an introduction that discusses the motivation of this work and a summary of related research, the chapter addresses the following topics:

- Demonstration that physisorbed gas molecules can localize excitons at low temperature and that laser irradiation can break the binding of physisorption.

- Analysis of the localization induced by gas molecules and explanation of the exciton redshift in temperature and blueshift with laser irradiation in terms of hopping.

- The effects of laser irradiation from charging of the monolayer to creation of defects.

Most of the essential contents of this chapter are published here [96].

3.1. Introduction

As discussed in Section 1.1, 2D semiconductors showed promising properties for new semiconductor technology. The interest on this class of materials is strictly connected to the atomic vertical scale. This property is relevant not only to scale down the dimension of opto-electronic devices but also for physical properties as for instance the inversion symmetry breaking. However, in order to take advantage of these opportunities in real-life applications, there are practical challenges that have to be tackled [97, 98]. Aspects that have to be considered are the long-term stability in ambient condition and the interaction between air molecules and the monolayers [10, 35, 99].

In fact, a prominent effect related to the extreme surface-to-volume ratio is the strong influence of physisorption and chemisorption of gas molecules on the optical and electronic properties [100, 101]. This efficient interaction has been exploited to engineer the materials for related applications, for instance chemical treatment for enhancing and tuning photoluminescence emission [102–104] as well as gas sensing and biosensing using modulations in electronic and optical responses as fingerprints for physisorbed gas molecules [105–107]. There is also a proposal and an experimental demonstration that PL can be used as a low-power consuming and contactless tool for gas sensing [108, 109].

Furthermore, physisorption is strongly connected with other prominent effects on TMD monolayers i.e. photo-annealing and photo-doping [110–112]. Recently, it has been shown that the exposure to visible light changes significantly the optical and electronic response of these materials. On one hand, these effects have to be kept under control for research investigations but on the other hand they can be used for technological applications such as reversible optical control of the doping [113] or laser micro-patterning [114].

For these reasons, an investigation on exciton localization by adsorbed molecules was carried out and will be presented in the next section. The research is conducted on $MoSe_2$ monolayers that is a paradigmatic compound among TMD monolayers. This material has similar optical properties as the other TMD compounds but it has a cleaner spectrum. Moreover, the possibility to use $MoSe_2$ monolayers as gas sensor both electronically [109, 115] and optically via PL [116] has recently been demonstrated experimentally.

3.2. Exciton localized by adsorbed gas molecules

The fabrication of the samples presented in this chapter was carried out by mechanical exfoliation and deterministic transfer via PDMS foils, as described in Section 2.1. Low-temperature PL was carried out using a commercial micro-PL setup coupled with a cryostat (Sec. 2.2.1). The excitation wavelength is

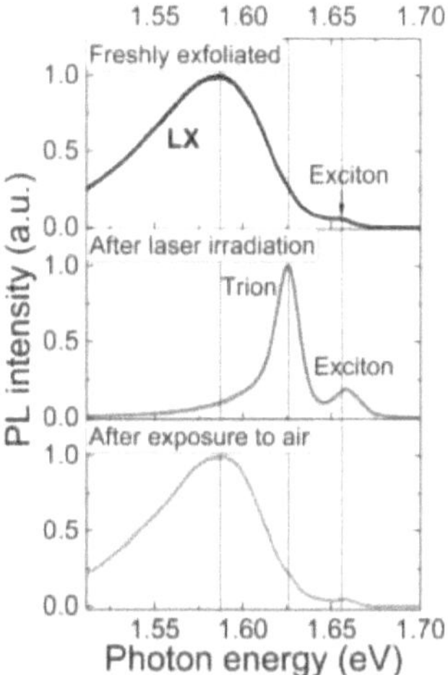

Figure 3.1.: PL of MoSe$_2$ monolayer at $T = 4$ K before and after laser-irradiation, and after a brief exposure to air. Before laser irradiation the localized exciton peak dominates the optical spectrum. After laser irradiation the trion peak dominates the optical spectrum. The changes induced by laser irradiation are completely reversible in air.

$\lambda = 532$ nm. The spot diameter on the sample is approximately 3 μm.

Excitons can be localized by adsorbed gas molecules. A straightforward proof of this localization is shown in Figure 3.1. The three spectra were measured using very low excitation power (100 nW) at $T = 4$ K. With this experimental conditions, localized exciton (LX) emission is the most efficient radiative channel since all the excited carriers relax towards the lowest energy states. In fact, in the panel above a broad emission is observed peaked at 1.58 eV and a tiny peak is observed at 1.66 eV. The second peak corresponds to free exciton emission in agreement with previous studies [117]. The broad peak at lower energy is a localized exciton peak as shown in literature [110]. Keeping the sample in vacuum and after laser irradiation at $\lambda = 532$ nm wavelength and with 0.5 kWcm^{-2} power density, the localized exciton peak almost disappears and the trion peak dominates the spectrum. These two effects are completely reversible after the exposure to air (third panel in Figure 3.1).

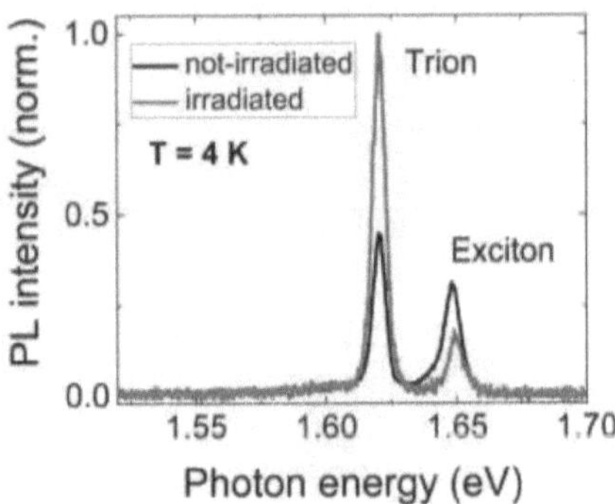

Figure 3.2.: PL spectrum at 4 K of MoSe$_2$ monolayer top-encapsulated in hBN. The photo-gating effect is observed, while there is no localized exciton peak. The spectrum of the irradiated sample was normalized and the spectrum of the non-irradiated sample was multiplied by the same factor.

This fact means that the low-energy peak is related to adsorption of gas molecules on the MoSe$_2$ surface and more specifically it points towards physisorption rather than chemisorption even if we cannot exclude that chemisorption plays any role [118]. The PL yield of the non-irradiated sample is around ten times higher than for the sample after irradiation due to the giant oscillator strength of the localized excitons, as is typical in semiconductors.

On the other side, the appearance of the trion peak is evidence for the increased number of available free carriers in the flake [119–121]. There is a double origin: both the desorption of physisorbed and chemisorbed polar and non-polar gas molecules (O$_2$ or H$_2$O) and a laser-induced spatial separation of carriers trapped into the substrate can generate a gating effect [100, 122]. In Figure 3.2, spectra of MoSe$_2$ monolayer top-encapsulated in hBN are shown. The absence of a localized exciton band demonstrates that the localization of excitons comes entirely from adsorbed gas molecules on the surface. Furthermore, even though the hBN prevents the localization of excitons by gas physisorbed gas molecules, optical illumination still changes the relative strengths of the trion and exciton signals. This gating effect is therefore due exclusively to the substrate gating.

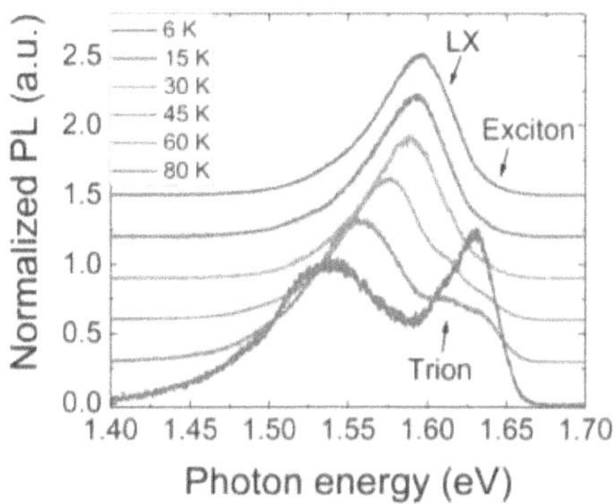

Figure 3.3.: Low-temperature PL spectra for different temperatures normalized with respect to the localized exciton peak. Curves are shifted vertically for clarity.

Another effect of laser irradiation is the slight blueshift of the exciton and trion peaks (Figure 3.2). The blueshift is smaller for the trion than for the exciton. The origin is a trade-off between increased dielectric screening due to the higher free electron population that reduces the exciton and trion binding energies, and a trion-exciton energetic distancing due to the increased Fermi energy [120,123].

3.3. Hopping between localization centers

Using extremely low excitation power, it is possible to avoid any modification in the physical properties of a freshly exfoliated sample. In this condition, temperature dependence PL was studied.

Figure 3.3 shows the low-temperature PL spectra for different temperatures. At low temperature most of the PL emission comes from the localized exciton states. The free exciton peak is barely visible. With increasing temperature the localized exciton peak drops in intensity and the exciton and the trion peaks appear. Nevertheless, the PL spectra are dominated till around 80 K by localized exciton emission. The shape of the localized exciton peak does not change much with temperature but there is a pronounced redshift and broadening.

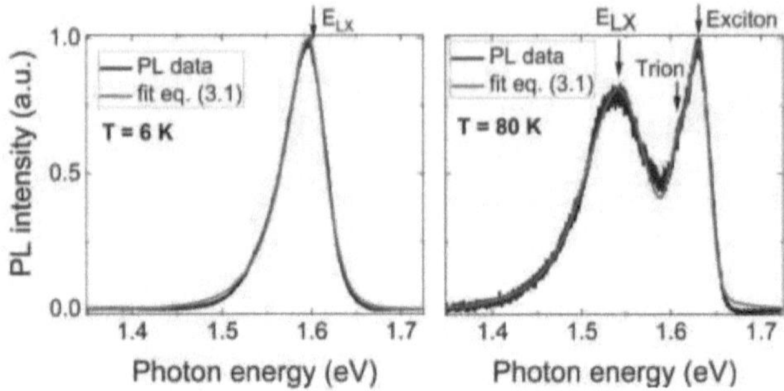

Figure 3.4.: Low-temperature PL spectra (black) and fit functions (red) at $T = 6$ K and $T = 80$ K. The fit functions are given in Equation (3.1).

In order to fit the lineshapes of the PL peaks, an analytical model for localized excitons was used. Similar models were used in previous studies [124–126]. This model takes into account an exponential sub-band-gap tail of states and a finite temperature that accounts for the thermal probability to escape from a localization center:

$$P(E) = \frac{I(T)\exp\left(\frac{E}{E_\mathrm{u}}\right)}{1 + \exp\left(\frac{E - E_\mathrm{LX}(T)}{k_\mathrm{B}T_\mathrm{PL}(T)}\right)} \tag{3.1}$$

where P is the PL intensity for every photon energy E, I is a normalization factor, E_u is the depth of the band-gap tail (Urbach energy [127]), T is the lattice temperature, T_PL is an effective PL temperature, and E_LX almost corresponds to the peak energy position at the maximum ($E_{max} = E_{LX} + k_B T_{PL} \ln\left(\frac{k_B T_{PL}}{E_u - k_B T_{PL}}\right)$). The Urbach-like part of this function is responsible for the low-energy side of the localized exciton peak in the fit and the Fermi-Dirac-like part is responsible for the high energy side. The PL temperature determines the slope of the exponential decay on the high-energy side of the peak. For higher temperatures where the exciton and trion peaks are more pronounced, two Gaussian functions centered at the exciton and trion resonances were used.

The fact that the localized exciton peak shows such a smooth distribution means that the density of physisorbed molecules is very high: it is possible that there is a full covering of gas molecules on top of the surface [105].

The agreement of this fitting function with the data is excellent in spite of the

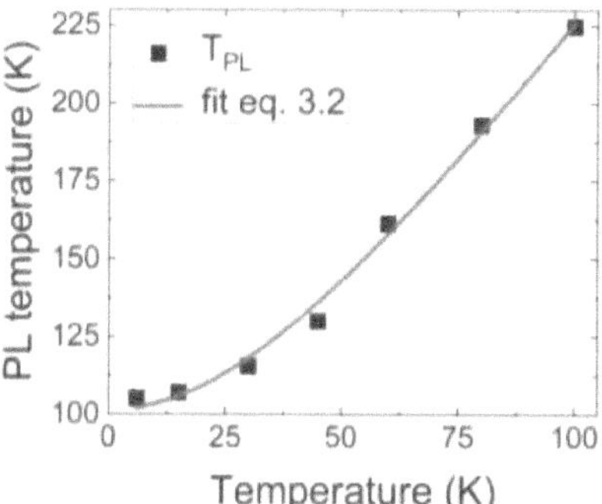

Figure 3.5.: Dependence of the effective PL temperature on lattice temperature. The fit function is given in equation (3.2).

small number of parameters as shown in Figure 3.4(a) and (b) at $T = 6$ K and $T = 80$ K, respectively. However, even for the lowest temperature we obtain from the fit PL temperature values above 100 K (see Figure 3.5).

From the data, E_u is independent of temperature - within our range of uncertainty - and has a value around 30 meV (Figure 3.6). An indication of this is already visible in the normalized spectra in Figure 3.3, where the exponential low-energy tail of the spectra does not change shape with temperature.

Other sub-band-gap density of states distributions rather than exponential have been tried out, e.g. Gaussian distribution or a Franz-Keldysh tail, but they result in worse agreement with the measured data (not shown) [128–130].

Figure 3.5 shows the dependence of the effective PL temperature on the actual lattice temperature. In analogy to Marianer et al. [131], the PL temperature can be interpreted as a pythagorean sum of the lattice temperature and an induced temperature T_0 at zero Kelvin:

$$T_{\mathrm{PL}}(T) = \sqrt{(\alpha T)^2 + T_0^2} \tag{3.2}$$

where α is a fitting parameter that for a direct band-gap semiconductor with parabolic dispersion should be identical to 1. As it is clear from Figure 3.5, the agreement with the experimental data is excellent, however $\alpha = 2.0 \pm 0.1$. The

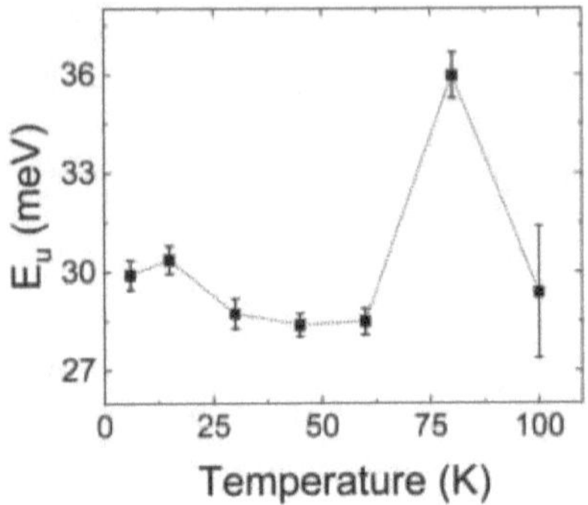

Figure 3.6.: Dependence of the Urbach energy E_u on lattice temperature. No clear trend is observed within the experimental error.

reason of the deviation of the value α from 1 is that the defect density of states was assumed to be exponential. Therefore, the effective PL temperature has to compensate artificially the non-parabolic dispersion with energy.

From the fit, $T_0 = 101 \pm 4$ K that corresponds to the energy of 9 meV (Figure 3.5). This value is interpreted as the quadratic deviation of the inhomogeneous potential distribution in the monolayer induced by physisorbed gas molecules. The induced strong inhomogeneity of potential has the effect to smear the sharp edge of the Fermi-Dirac part of Equation (3.1). The possible effect of higher carrier temperature with respect to the lattice temperature was considered as well. However, this interpretation was ruled out since extremely low excitation power was used throughout all the experiment (100 nW).

The energy E_{LX} has a clear temperature dependence (Figure 3.7). In this figure, the contribution of the redshift due to the renormalization of the band gap was already subtracted using the Varshni formula [117]. This correction is anyway almost one order of magnitude less than the observed redshift in this temperature range.

The redshift of the localized exciton peak with temperature is clearly not only due to the redshift of the renormalization of the band gap, but there is another physical effect involved. This effect could be related to hopping between localized states [132, 133]. Hopping between defects means that a localized exciton can

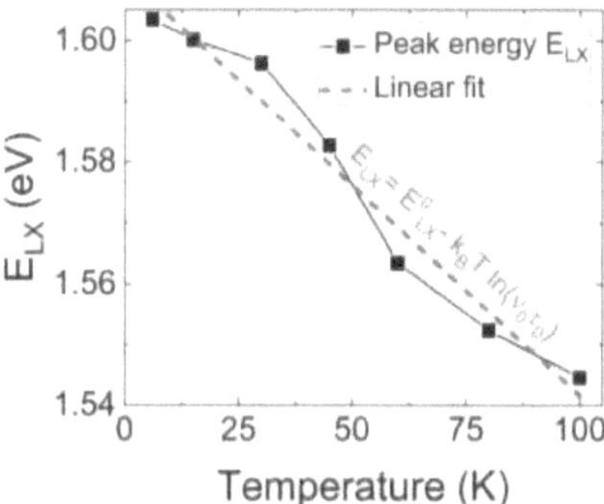

Figure 3.7.: Temperature dependence of the localized exciton peak energy E_{LX}. A monotonic decrease of the peak energy is observed and is explained in terms of hopping between localization centers.

anyway travel across the flake through hopping between different localization centers. Furthermore, excitons prefer to hop into a favorable lower energy state (infrared side). Increasing the temperature means also increasing the probability for hopping (mobility of a localized exciton) that leads to a redshift of the localized exciton peak. This effect has already been shown for different semiconductors with reduced dimensions like diluted nitride quantum wells [134] and quantum dots [135].

An analytical model that quantifies the redshift with temperature due to the hopping effect has been developed by Baranovskii et al. [136] for an exponential sub-band-gap density of states. The peak energy can be approximated by $E_{\mathrm{LX}} = E^0_{\mathrm{LX}} - k_\mathrm{B} T \ln(\nu_0 \tau_0)$ where ν_0 is the attempt-to-escape rate from a defect and τ_0 is the lifetime of a localized exciton at $T = 0$ K. From the linear fit in Figure 3.7, we get $\ln(\nu_0 \tau_0) = 8.1$. Typically, ν_0 is of the order of 10^{12} s^{-1} [132], therefore the localized exciton lifetime is of the order of 5 ns that is in good agreement with experimental data [137].

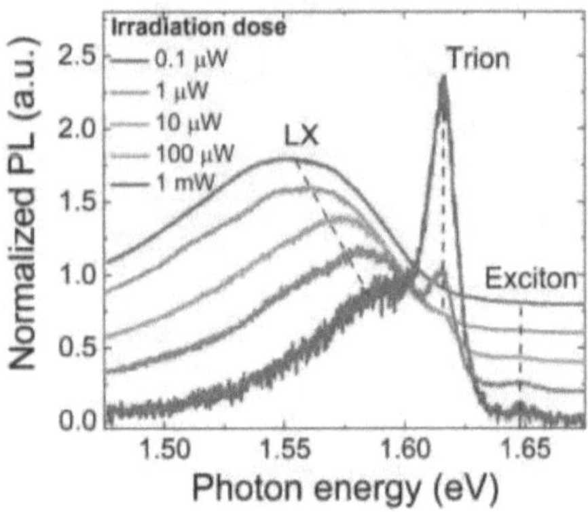

Figure 3.8.: Normalized PL spectra at $T = 5$ K for different irradiation doses and same excitation power (100 nW). The sample were irradiaed for around 60 sec at each power and the spot diameter of the laser was 3 μm. Curves are shifted vertically for clarity. A monotonic blueshift of the localized exciton peak is observed at higher irradiation doses.

3.3.1. Hopping and laser annealing

Using the same approach, the analysis of the effects of laser irradiation on the localized exciton peak is presented. The spectra are measured using low excitation power (100 nW) after different laser irradiation doses (Figure 3.8). Again the exponential tail does not change shape up to hundreds of μW of laser power, while the PL temperature decreases and the peak energy shows a blueshift. According to the previous interpretation, the blueshift is due to a transfer of population rather than an energetic redistribution of defect states. More specifically, the hopping probability drops due to the lower amount of gas molecules available as localization sites on the surface after the annealing. Consistently, the effective PL temperature decreases because the disorder potential induced by the gas molecules is smaller.

Above 1 mW of irradiation power, a decrease of the trion peak PL intensity was observed (not shown) which is a consequence of an irreversible deterioration of the crystal quality. Simultaneously, the shape of the localized exciton peak starts to change.

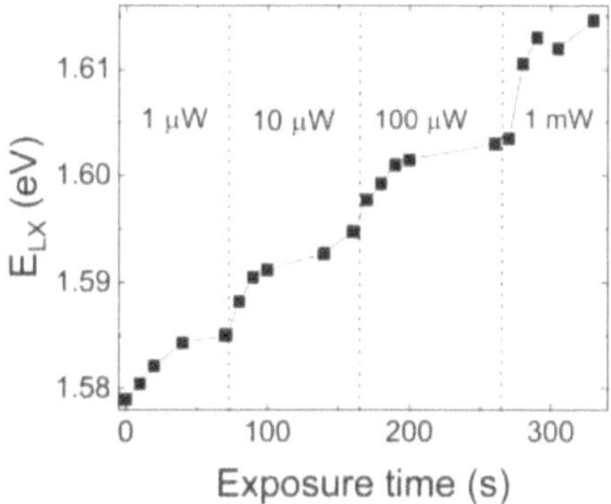

Figure 3.9.: Evolution in time of the peak energy emission E_{LX} for different irradiation doses (same excitation power and temperature).

Figure 3.9 shows the energy shift of the localized exciton peak with increasing dose of laser exposure. At a fixed irradiation exposure, a monotonic exponential blueshift of the peak energy is observed and it saturates in a time scale of minutes. This saturation behavior is due to the finite number of physisorbed gas molecules that is possible to remove with a given irradiation dose. At a different power density, the shift of the peak enters into an another regime and new exponential change in the energy position of the localized exciton is observed.

A good experimental procedure to keep the TMD samples unchanged during an optical measurement is to irradiate the sample at higher power for few seconds before measuring at the lower power desired for the real measurement.

Figure 3.10 shows the dependence of the parameter T_{PL} at 6 K on laser doses. As discussed above, this parameter takes into account the disorder potential in the monolayer. After laser irradiation, i.e. after desorption of gas molecules, the PL temperature decreases. This indicates that gas molecules physisorbed on the surface are one of the most important factor that induces a disorder in the material. Therefore, laser irradiation works as a laser annealing of the crystal.

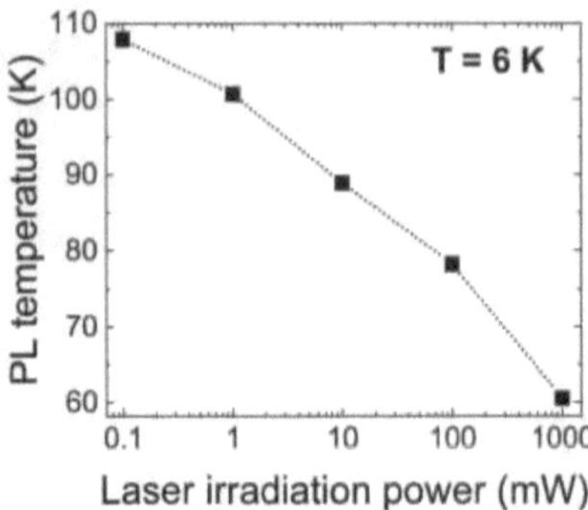

Figure 3.10.: Dependence of the effective PL temperature on the laser irradiation. A strong decrease of the PL temperature suggests a decreases of the disorder potential in the monolayer after laser irradiation.

3.4. Effects of laser irradiation: dynamics

In the previous sections, the effects of laser irradiation on the optical properties of $MoSe_2$ monolayer were discussed and analyzed. In this section, the effects of laser irradiation on the dynamics of radiative recombination are presented.

In order to study the effects on the dynamics, time-resolved PL was measured. The second harmonic of a mode-locked Titanium:sapphire laser system was used as excitation source ($\lambda = 400$ nm). The PL emission is dispersed into a spectrometer and then coupled into a streak tube and a CCD camera. The setup used for this measurement is described in Section 2.2.2.

Figure 3.11(a) shows the time-resolved PL spectra of a freshly exfoliated sample. The trion and exciton peaks are observed and they show a fast exponential decay. The exciton lifetime is shorter that the time resolution of the detection system ($\tau_X \leq 2$ ps). The trion has a lifetime of $\tau_{Tr} = 13 \pm 2$ ps in agreement with previous studies [138]. The PL decays of the two excitonic resonances are shown in Figure 3.11(b).

The power density used for the measurement is around 5 J cm^{-2}. This value is enough to remove all the physisorbed gas molecules on the surface and, therefore, no localized exciton peak is observed. Furthermore, the lifetime of localized

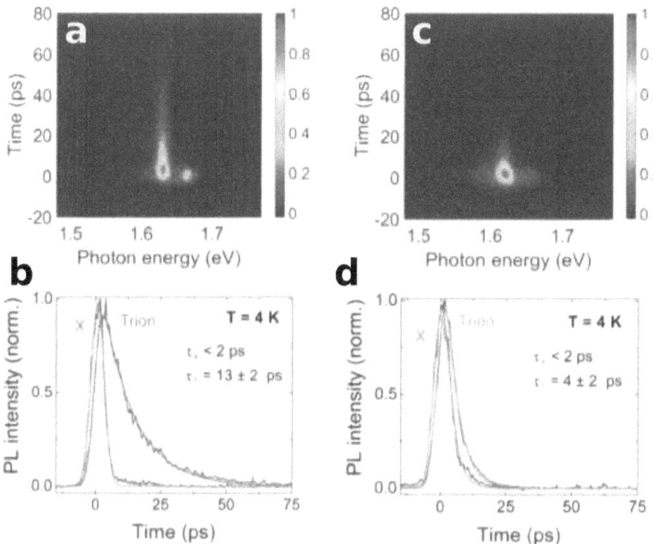

Figure 3.11.: (a) and (c) false-color 2D map of the time-resolved PL at $T = 4$ K before and after laser irradiation. (c) and (d) PL decays of exciton and trion before and after irradiation with a mono-exponential fit function.

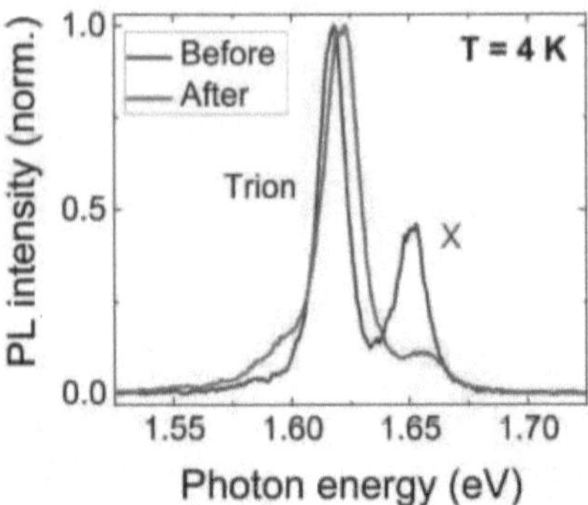

Figure 3.12.: Time-integrated spectra before and after laser irradiation. A photo-doping is observed with a blueshift of both the trion and the exciton resonances. The localized exciton peak is not visible since the experiment probed the first 80 ps after the NIR pulse arrival.

exciton is on the order of tens or hundreds of nanoseconds [137]. Therefore, the PL emission due to the localized exciton is not observable with this time scale.

After intense laser irradiation with a pulsed laser (25 μJ cm^{-2} for around 100 seconds), the PL dynamics changes significantly (Figure 3.11(c)). The trion lifetime gets shorter ($\tau_{Tr}^{Irr} = 4 \pm 2$ ps) (Figure 3.11(d)). At the same time, the relative weight of exciton and trion intensity changed indicating a photo-doping effect (Figure 3.12). The linewidths of the exciton and the trion peaks increase, i.e. inhomogeneous broadening. Simultaneously, a tail appears at lower energies. These evidences indicate that the this laser fluence is sufficient to induce permanent changes in the crystal, i.e. creation of crystal defects. The trion lifetime is therefore dominated by non-radiative defect scattering.

These results suggest that laser irradiation could be an useful tool for tailoring the optical properties of monolayer with micrometric resolution. Furthermore, the effects of laser irradiation have to be always taken into account when an experiment is performed, especially when short laser pulses are used.

3.5. Conclusions and outlook

In summary, an investigation of the optical properties at low temperature of monolayer $MoSe_2$ was presented in this chapter. In particular, it has been shown that adsorbates can localize excitons at low temperatures. Irradiation with a laser at low power is sufficient to release the adsorbates and modify the PL emission.

Upon increasing the temperature, a strong redshift of the localized exciton peak was observed. Complementary, a blueshift of the same peak was observed after laser irradiation. Both energy shifts were explained in terms of exciton hopping between localization centers.

Finally, the effects of laser irradiation on the PL dynamics were shown. After laser irradiation, a decrease of the trion lifetime was observed, together with a broadening of the excitonic lines.

The work presented in this chapter is relevant both for the fundamental understanding of the optical properties of this class of semiconductors but also for more practical purposes. In fact, the research can be useful for the realization of gas sensors based on TMD monolayers. Furthermore, this study is of direct interest for all the optical experiments that will be presented in the next chapters, since laser irradiation has such strong effects on the optical properties of all the TMD monolayer and, more in general, of two-dimensional materials.

An on-going project which started after the publication of this research is the collaboration with Dr. Christian Wagner (HZDR) and Prof. Angela Thränhardt (TU Chemnitz). In this project, numerical simulations based on Monte Carlo methods are performed considering hopping as the mechanism for the energy shifts of the localized exciton peak observed in the experiments. With the help of the simulations and with the direct comparison with the experiments, it is possible to get a quantification of the hopping parameters with the final aim to get a realistic quantification of the number of gas molecules on the monolayer surface just looking at the PL spectra.

4. Infrared response of TMD monolayers

During the course of this PhD, several experiments on TMD monolayers were performed using the infrared free-electron laser at FELBE. These experiments are presented in this chapter that is divided as follows:

- An introduction on the infrared properties of TMD monolayers and the main achievements of the latest research are discussed.

- An experiment of infrared-pump and optical-probe spectroscopy will be presented. This experiment shows that the trion resonance exhibits a redshift when non-resonant infrared radiation is applied as a consequence of free carrier absorption.

- Two other FEL experiments will be shortly presented. The first is about intraexcitonic transitions and the second is about surface phonon-polaritons. These two experiments did not give the expected outcomes and the reasons will be shortly discussed.

4.1. Infrared properties of TMD monolayers

Most of infrared technology is based on semiconductors and their heterostructures. Van der Waals semiconductors are an emerging class of materials and are in the focus of intense research. However, only few experiments have been performed for investigating the infrared properties and THz applications of TMD monolayers. The reason is the extremely low absorption of terahertz radiation. The losses are so low that it has been proposed to use TMD monolayers as transparent electrodes for THz applications [139]. The low absorption is due to the extremely small thickness, i.e. more than three orders of magnitude smaller than the wavelength of the radiation.

However, when the carrier concentration in the monolayer is sufficiently high, the monolayer becomes more sensitive to the THz radiation because of free carrier absorption and excitation of plasma waves. Remarkably, a THz photodetector based on $MoSe_2$ monolayer has been fabricated and measured, even if the performances are mediocre [140].

In order to enhance the interaction between TMD monolayers and infrared and THz radiation, plasmonic structures or metamaterial systems such as photonic crystals can be used, as was demonstrated for graphene [141,142]. However, there are no reports of such devices working in the infrared range for TMD monolayers or TMD heterostructures.

Another technological issue for THz applications and research is the growth of large-area TMD monolayers. The first demonstration of a large-area TMD monolayer was in 2014 and many progresses have been achieved until now [143]. However, the crystal quality is still not comparable with exfoliated samples and the carrier mobility is orders of magnitude lower than predicted by theoretical calculations. The size of exfoliated flakes is on the order of tens of microns. Therefore, the spot diameter of focused THz radiation is usually considerably larger than the flakes. This complicates and slows down considerably the development of the research on this topic.

Nevertheless, interesting experiments that use infrared radiation have been already carried out. Optical-pump THz-probe spectroscopy has been used in different variations mostly to investigate the induced photo-conductivity of TMD monolayers. A very fast response on the order of hundreds of femtoseconds has been measured at room temperature [144]. Interestingly, a negative photo-conductivity has been observed at low temperature in intrinsically doped MoS_2 monolayer [145]. This is tentatively attributed to the formation of trions that reduce the number of free carriers and create quasiparticles with large effective mass.

With a similar experimental scheme, the internal exciton transitions (1s-2p and so on) have been directly measured with broadband mid-infrared radiation after the creation of an exciton population with optical pulses [19]. In fact, since the exciton binding energy is very high (hundreds of meV), the internal transitions fall in the MIR range (e.g. $\Delta E_{1s-2p} = 140$ meV in WSe$_2$ monolayer).

Intense mid-infrared pulses have been also used for high-order exciton sideband generation in WSe$_2$ monolayer [146]. However, to get in this regime, peak field strength higher than 1 MV cm^{-1} was needed. This indicates that bare TMD monolayers are not very sensitive to THz and infrared radiation.

Recently, intersubband transitions in thin flakes of WSe$_2$ were observed [147]. In fact, any van der Waals thin layer forms a natural quantum well with the barriers given by the vacuum. Similar and even more promising results were found in InSe [148, 149]. These two latter experiments are very promising for the development of THz and infrared applications based on two-dimensional materials.

One of the main topic of this PhD is the investigation of the infrared response of TMD monolayers by means of the infrared FEL. Different experiments were performed and they will be presented in next sections.

4.2. Trion redshift induced by THz radiation in MoSe$_2$ monolayer

As discussed above, the use of two-dimensional semiconductors for infrared applications is still elusive. At the same time, the research is not proceeding very fast in this direction: only few pioneering experiments have used infrared radiation in order to manipulate the properties of two-dimensional materials. There is still a lot of room for scientific research and technological development.

In this experiment, the influence of intense infrared radiation on the optical properties of MoSe$_2$ monolayers is investigated. For this purpose, a THz-pump and NIR-probe experiment was carried out. For the pump beam, the free electron laser FELBE is used. The characteristics of this source are described in detail above (Section 2.3). As probe beam, a narrowband Titanium:Sapphire oscillator is used. The probe wavelength is tuned across the exciton and trion resonances and the reflectivity of the samples is measured. The experimental setup is sketched in Figure 4.1.

The sample is fabricated via mechanical exfoliation as described in the previous section (Sec. 2.1). The substrate is high-resistivity silicon with a thin layer of 90 nm SiO$_2$ on top. With this substrate, the absorption of THz radiation by the

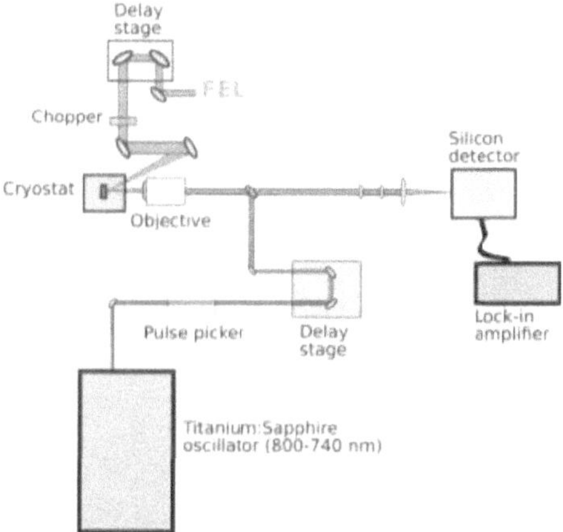

Figure 4.1.: Sketch of the THz-pump NIR-probe setup.

substrate is very low and, at the same time, good optical contrast is assured by the thin layer of SiO_2.

The pump-probe setup is arranged in a reflection geometry and the measurements are carried out at low temperatures. In order to characterize the samples, reflection spectra of $MoSe_2$ monolayer at low temperature have been measured and are shown in Figure 4.2. The spectra are measured with an ultrafast broadband near-infrared source with a pulse length of 10 fs. The spectra are measured after different laser irradiation fluences. The irradiation was done with the same laser used for the pump-probe experiment, i.e. narrowband Ti:Sapph oscillator, at 1.650 eV photon energy in resonance with the exciton energy. This was done because even a low-power near-infrared pulse changes the optical properties of the TMD monolayers and the effects of the NIR probe beam have to be kept under control [122].

The reflection is defined by the reflection spectrum of the monolayer on top of the substrate divided by the spectrum of the source.

With increasing fluences, the trion resonance appears as consequence of photodoping caused by the desorption of adsorbates on the surface and a charging of the substrate [96,120]. The changes in the reflection spectra are due to permanent changes in the samples as described in the previous chapter.

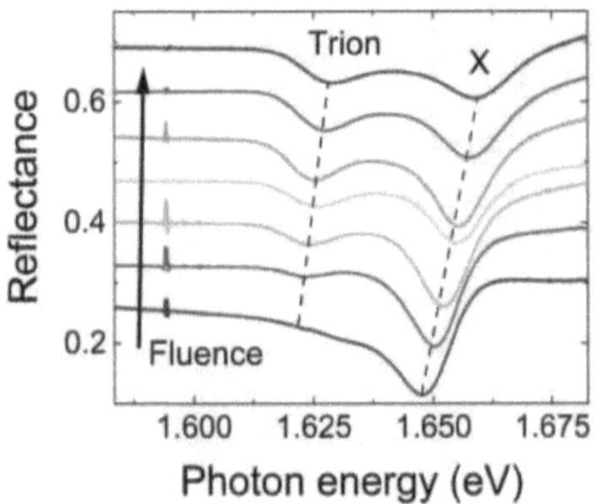

Figure 4.2.: Reflection spectra of MoSe$_2$ monolayer at 5 K. After laser irradiation (around 5 minutes at 1.650 eV photon energy), the trion resonance appears because of photo-doping. At higher fluences, the trion and the exciton resonances exhibit a blueshift and broadening. The fluences per pulse are 0.2, 2, 5, 14, 29, and 60 μJ cm^{-2}. The spectra are vertically displaced for clarity.

A blueshift of the trion and exciton resonances is observed at higher fluences and, simultaneously, the energetic distance between the trion and the exciton resonances increases. These effects indicate that the free carrier concentration in the sample, i.e. the doping, increases. In fact, the energetic distance between trion and exciton energy depends on the Fermi energy:

$$\hbar\omega_X - \hbar\omega_{Trion} = E_{Trion} + \epsilon_F \tag{4.1}$$

where E_{Trion} is the trion binding energy and ϵ_F is the Fermi energy with respect to the bottom of the conduction band [123].

From Equation (4.1) and from the comparison with literature, a free electron concentration of around 10^{12} cm^{-2} is estimated for the samples studied in the pump-probe experiments [150].

4.2.1. FEL-pump NIR-probe results

In order to keep under control the carrier concentration in the sample during the pump-probe measurements, a relatively low NIR probe fluence is used and the reflection spectra are measured continuously. Furthermore, the sample was first irradiated with a fluence of 15 μJ cm^{-2} and then measured with a lower NIR probe fluence. This assures no significant changes in the samples during the experiments.

The infrared spot diameter was measured to be 400 μm, i.e. much larger of the MoSe$_2$ monolayer flake. However, the NIR probe laser was focused on a spot diameter of 3 μm. This gives the spatial resolution of the experiments.

Figure 4.3 shows the evolution in time of the differential reflection spectrum at 8 K. The FEL excitation energy is 32 meV, far below any electronic transition and not in resonance with the infrared active phonon energies. The NIR narrowband probe is tuned across the exciton and trion resonances in order to measure the changes in reflection induced by the THz pump.

A strong signal at the trion energy dominates the differential reflection spectrum. A weaker signal at the exciton energy is also observed. Furthermore, the signal at the trion energy shows a slow component that is not observed at the exciton energy. The two features have a dispersive shape that indicates a redshift

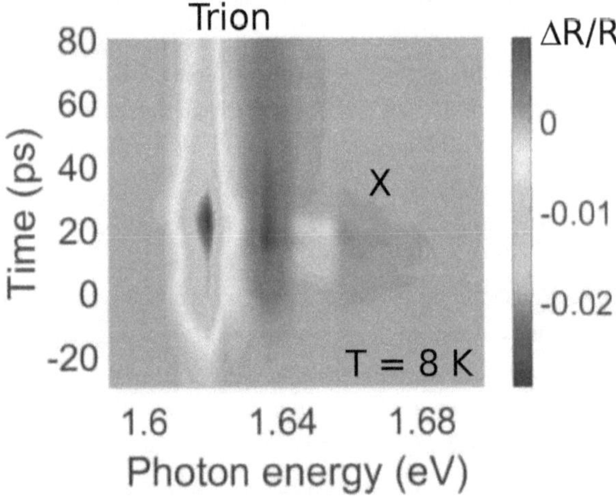

Figure 4.3.: Differential reflection spectra in time at $T = 8$ K. A strong and long-living feature is observed at the trion energy. At the exciton energy a fast and weak signal is also observed. The FEL fluence is $\phi_{FEL} = 5\ \frac{\mu J}{cm^2}$ at 32 meV photon energy.

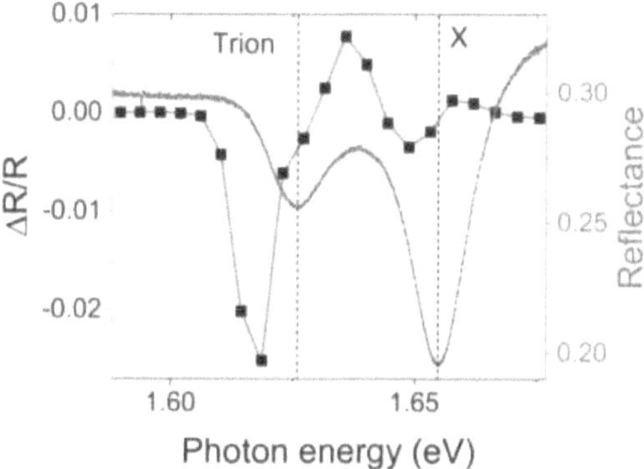

Figure 4.4.: Differential reflection spectra at the time overlap in comparison with the linear reflectivity. The dispersive shapes of the differential signal are located at the trion and exciton energies.

of the two resonances induced by the infrared radiation.

In order to understand better the sign and the spectral dependence of the differential reflection spectra, Figure 4.4 shows the differential spectra at the time overlap of the pump and probe pulses in comparison with the linear reflectivity. The zero-crossing points of the dispersive shapes of the differential signals are located at the trion and exciton energies. The sign of the signal changes at each resonance.

The weak signal at the exciton energy shows an instantaneous decay, i.e. the decay constant is shorter than the time resolution of the setup (Fig. 4.5(a)). The time resolution is basically limited by the pulse length of the FEL. The pulse length of the FEL is $\tau_{FEL} = 5.5$ ps as it is estimated from the Fourier transform of its spectrum (see appendix A.3). The convolution of the FEL and the probe pulses gives the time resolution that is $\sigma_\tau = \sqrt{\tau_{FEL}^2 + \tau_{NIR}^2} = 6.2$ ps. The pulse length of the probe beam is measured with an autocorrelator and is around $\tau_{NIR} = 3$ ps.

Figure 4.5(a) shows clearly that two pump pulses arrive at the sample. This is due to the reflection at the back side of the substrate. The temporal distance between the two pulses is around 15 ps which corresponds to a distance of $d = \frac{c\Delta t}{n_{Si}} = 1.3$ mm, about twice the substrate thickness. The arrival of more than one

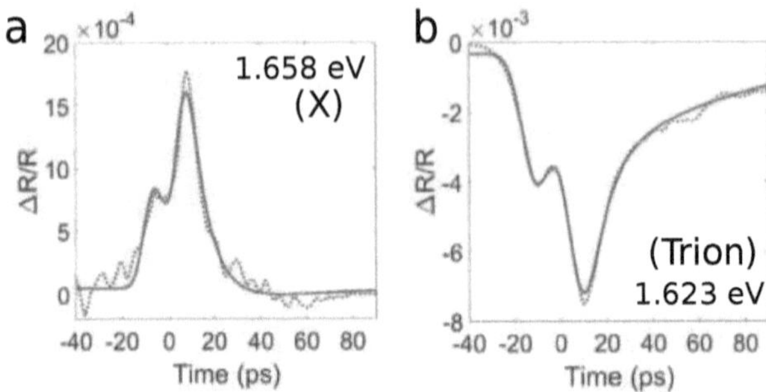

Figure 4.5.: (a) and (b) Differential reflection decays at 1.658 eV and 1.623 eV probe energy, respectively. The fast component is limited by the convolution of the pump and probe pulses. The slow component is present only at trion energies and has a decay constant of $\tau_2 = 70 \pm 10$ ps.

pump pulse is a very common issue for infrared spectroscopy. This fact has to be taken into account for the modeling of the instrument response function that is now a sum of two Gaussian with same broadening but different central time and intensity.

The instantaneous differential signal at the exciton is present only when pump and probe pulses are overlapping in time on the sample. Therefore, this signal can be due to an optical Stark redshift of the exciton, as observed by Yong et al. when the pumping radiation energy is tuned below the 1s-2p transition [151].

The signal at the trion energy shows a bi-exponential decay. The decay of the differential reflection is shown in Figure 4.5(b). The fast component is again present and the two pump pulses are observed. However, besides the fast component of the decay, a slow component is observed. In order to fit the data and to get the decay constants, a bi-exponential decay is used for signal at the trion energy. To fit the data at the exciton energy, an exponential decay function with decay constant smaller than the pulse length is used.

The slow component at the trion resonance has a lifetime of $\tau_2 = 70 \pm 10$ ps. The fast components are shorter than the time resolution $t^{fast}_{X,Trion} \leq 6$ ps.

A bi-exponential decay function can reproduce the experimental data with good agreement. By fitting the decays at every photon energy, it is possible to isolate

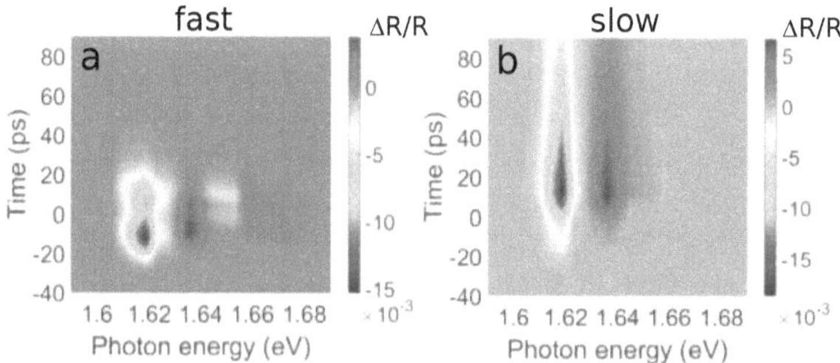

Figure 4.6.: (a) and (b) Fast and the slow components of the differential reflection as obtained from a bi-exponential fit.

the fast and the slow components of the decay. Figure 4.6 shows a 2D false-color map of the two decay channels separately. The decay constants τ_1 and τ_2 are not fixed. However, they do not vary significantly for different photon energies.

Similar results were obtained performing the experiment at different FEL photon energies. The experiment was repeated at 14 meV and 24 meV (not shown). This demonstrates that the signal comes from a non-resonant effect.

4.2.2. Discussion

As mentioned above, the fast exciton decay is consistent with an exciton AC Stark shift as reported previously in $MoSe_2$ monolayer [151]. Differently from an exciton, a trion resonance does not admit any internal excited bound states [152]. But, if the exciton population gets a redshift in energy after the photon dressing (exciton AC Stark), the trion energy is also shifted. However, this cannot explain the large trion redshift. In other words, the differential reflection signal at the trion energy would be expected to be smaller than the one at the exciton energy because the oscillator strength of the trion resonance is smaller, as it can be observed in the linear reflection spectrum of figure 4.2.

Furthermore, a modulation of the signal at the trion energy is observed also when the infrared and the probe pulses are not overlapping in time on the sample. This implies that there is an absorption of infrared radiation and a subsequent relaxation. The absorption mechanism is free carrier absorption leading to carrier heating.

Other absorption mechanisms have also been considered. Heating the mono-layer lattice via FEL absorption or via substrate heating are possible phenomena. However, the dynamics of such processes take place on a much longer time scale. Furthermore, this would not explain why the signal at the trion energy is around one order of magnitude higher than the signal at the exciton energy. Moreover, a slow component of the decay would have been expected both at exciton and the trion energy.

Another possible phenomenon is the temporary change in the refractive index of the substrate that would shift the band-gap energy. This is expected only at the time overlap and it cannot explain the slow component at the trion en-ergy. Furthermore, it cannot explain the huge difference of the intensities of the differential signal of trion and exciton.

Free carrier absorption gives an explanation of the redshift of the trion peak, while leaving the exciton peak almost unperturbed. Furthermore, the time scale of the cooling (tens of picoseconds) is in better agreement with electron gas heating and cooling. Intuitively, it is reasonable that the acceleration of the electron cloud has much stronger effect on the trion resonance. An intuitive explanation of the trion redshift induced by the free electron acceleration is that the absorption (or reflection) of a NIR photon that creates a trion requires less energy because the rest of the energy is taken by the kinetic energy of the accelerated electron gas. Therefore, the more the free electron gas is accelerated (hot electron gas), the less energy is required for the optical transition and a redshift of the trion peak is observed. Using the energy and momentum conservation, the formation of a trion has to obey to the following equations when the momentum of a free electron is transferred to the trion:

$$p_e = p_T, \tag{4.2}$$

$$\Delta E_T = E_T - E_e = \frac{p_T^2}{2M_T} - \frac{p_e^2}{2m_e} = \frac{p_e^2 M_x}{2m_e M_T} \tag{4.3}$$

where p_e and p_T are the electron and trion momenta, ΔE_T is the trion energy shift, m_e is the electron mass, $M_T = 2m_e + m_h$ and $M_X = m_e + m_h$ are the trion and exciton masses. Therefore, the formation of a trion starting with a high-momentum electron leads to a redshift ΔE_T of the trion resonance.

In order to get a deeper and quantitative understanding of this phenomenon, the exciton and trion Green functions are considered with the aim of getting an expression for the linear susceptibility as was done by Esser et al. [153]. The complete derivation of the Green functions and the susceptibility is reported in Appendix A.4. The idea is to get the microscopic polarization in the density matrix formalism and study its evolution in time by the Von Neumann equation $i\hbar\partial_t(P_X + P_T) = [H, P_X + P_T]$. By expressing the polarization in the frequency space via Fourier transform, the susceptibility is obtained from $\chi(\omega) \propto \sum_k P_k(\omega)$.

Considering only the ground state of the trion and the singlet state for its isospin, the linear susceptibility is:

$$\chi(\omega) = \chi_0\left((1 - N_e)\sum_n \frac{|\psi_n^2|}{E_n^X - \hbar\omega} + \frac{1}{\Omega}\sum_q f_q^e \frac{|M^T(q)|^2}{E^T - W_q - \hbar\omega}\right), \qquad (4.4)$$

where the first addend is related to the exciton states, N_e is the total carrier density, ψ_n is the excitonic wavefunction, E_n are the exciton energies, Ω is a normalization volume factor, q is the free carrier momentum, $M^T(q)$ is the oscillator strength, E^T is the trion energy, and $W_q = E_q^e - \frac{\hbar^2 q^2}{2M_T} = \frac{\hbar^2 q^2 M_X}{2m_e M_T}$ is the difference of the kinetic energy of the free electron and the center of mass of the trion (electron recoil) [153]. $M_T = 2m_e + m_h$ is the trion effective mass, and $M_X = m_e + m_h$ is the exciton effective mass, as described above.

As it becomes clear from Equation (4.4), if the free carrier that composes the trion had an initial momentum q and kinetic energy E_q^e, the energy and the momentum are transferred to the trion quasiparticle and the optical transition energy is redshifted by the factor W_q. Differently, the exciton transition energy is not influenced by the kinetic energy of the free carrier sea. The physical picture is sketched in Figure 4.7.

Besides a redshift of the peak, a low energy tail at the trion resonance appears as studied previously in MoS_2 monolayer and GaAs quantum wells [154, 155].

The exciton optical transition can occur only inside the light cone. Differently, a trion can relax even if it is outside the light cone. In fact, the momentum can be taken by the remaining free electrons.

In order to get the trion lineshape dependence on the carrier momentum, the

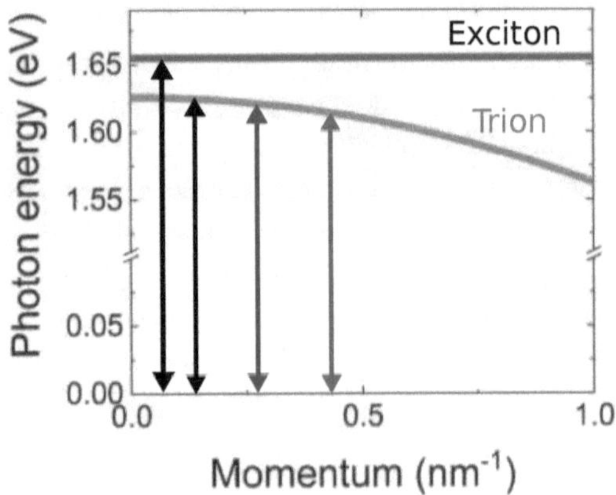

Figure 4.7.: Sketch of the optical trion and exciton transition as a function of carrier momentum. The initial kinetic energy of the free carrier is transferred to the trion, reducing the actual optical transition energy ($E_q^T = E_0^T - \frac{\hbar^2 q^2 M_X}{2 m_e M_T}$). The exciton transition energy is unperturbed.

oscillator strength $M^T(q)$ has to be calculated. The expression for the trion oscillator strength is:

$$M^T(q) = \frac{1}{\Omega} \sum_k \psi_T(k, q) = \int dr_2 \psi_T(r_1 = 0, r_2) e^{iqr_2 \frac{M_X}{M_T}}, \qquad (4.5)$$

where $\psi(r_1, r_2)$ is the trion wavefunction that depends on the relative position of the two electrons in the trion center of mass [153]. The integral has to be evaluated fixing the position of one electron at the origin, i.e. where the hole is, and let the other free. This is because one electron has to be at the position of the hole in order to make the optical transition.

The trion wavefunctions have to be calculated solving the three-body Schrödinger equation [153]. However, a variational form can be assumed as the symmetrized product of excitons wavefunctions with two different radii:

$$\psi_T(r_1, r_2, a, b) = \frac{\sqrt{2}}{\pi ab}(e^{-\frac{r_1}{a} - \frac{r_2}{b}} + e^{-\frac{r_2}{a} - \frac{r_1}{b}}) \qquad (4.6)$$

where a and b are the radii of the two exciton functions [156]. The two radii are necessary in order to get a bound state in the variational approach and it is important in order to account for the electron-electron repulsion. Equation (4.5) is basically the Fourier transform of the trion wavefunctions, i.e. exponential decay. Therefore, the oscillator strength M^T is a sum of two Lorentzian functions centered a $q = 0$:

$$M^T(q) = -\frac{4\sqrt{2}}{ab}\left(\frac{\frac{1}{a} + iq\frac{M_X}{M_T}}{\frac{1}{a^2} + q^2(\frac{M_X}{M_T})^2} + \frac{\frac{1}{b} + iq\frac{M_X}{M_T}}{\frac{1}{b^2} + q^2(\frac{M_X}{M_T})^2}\right). \tag{4.7}$$

The radii a and b are assumed to be 0.9 nm and 1.9 nm as calculated in literature [157]. The trion is quite localized in the real space, therefore it has a smooth distribution in the momentum space and the matrix element does not play a crucial role for the experimental data under consideration.

Assuming a Maxwell-Boltzmann distribution for the thermal free carrier distribution $f_q \propto q^2 \exp\left(-\frac{\hbar^2 q^2}{2m_e K_B T_c}\right)$ and the relations $R(\omega) \simeq -a(\omega) \propto \mathrm{Im}\,\chi(\omega)$, it is possible to get an expression for the reflectance as a function of free carrier temperature T_c. The expression for the reflectance is then:

$$R(\omega, T_c) = R_0 - c\,\mathrm{Im}\left((1-N_e)\sum_n \frac{|\psi_n^2(r=0)|}{E_n^X - \hbar\omega - i\delta_X} + p\sum_q f_q^e(T_c)\frac{|M^T(q)|^2}{E^T - W_q - \hbar\omega - i\delta_T}\right) \tag{4.8}$$

where δ_X and δ_T are introduced to take into account an inhomogeneous broadening of the exciton and trion resonances, respectively, and c and p are the empirical weights used to fit the measured spectra in Figure 4.4 with the calculated reflection spectra.

The reflection spectra as a function of carrier temperature are shown in Figure 4.8. To get the broadening $\delta_{X,T}$ and the intensities of the two resonances c and p, the reflection spectrum at 5 K shown in Figure 4.4 was used.

In the calculation above, the effect of the lattice temperature was not taken into account. It was assumed that the heat capacity of the free carrier system is much smaller than heat capacity of the lattice and negligible heat is transferred from the hot electron system to the lattice.

The infrared radiation in the experiment has the role to accelerate the free carrier population. Assuming a quasi-thermal equilibrium of the carriers at any delay time, the time evolution of the pump-probe signal can be reproduced as

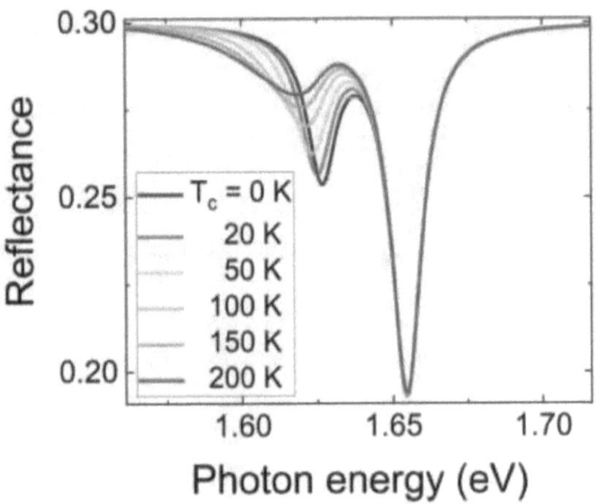

Figure 4.8.: Reflectance of MoSe$_2$ monolayer as a function of the free carrier temperature T_c. The lattice temperature is $T = 0$ K.

differential reflection:

$$\Delta R(\omega, t) = \Delta R(\omega, T_c) = \frac{R(\omega, T_c) - R(\omega, T_0)}{R(\omega, T_0)} \qquad (4.9)$$

where $R(\omega, T)$ is obtained from Equation (4.8). The differential reflection signal at the trion energy is then fitted with this function letting only the temperature of the electron gas changing in time. The calculated differential reflection in time is shown in Figure 4.9. The qualitative behavior of the experimental observation is well reproduced (see Figure 4.3).

The heating of the free carrier system reproduces also the decrease in intensity of reflection at the trion energy. In fact, at higher carrier temperature, the free carrier has a higher average momentum. However, the matrix element $M^T(q)$ decreases at higher q or, in other words, the oscillator strength of the trion resonance decreases at larger momenta.

However, the shape of the signal could not be perfectly reproduced. Figure 4.10 shows the comparison between the measured and the calculated differential reflection spectrum at a given time (70 ps).

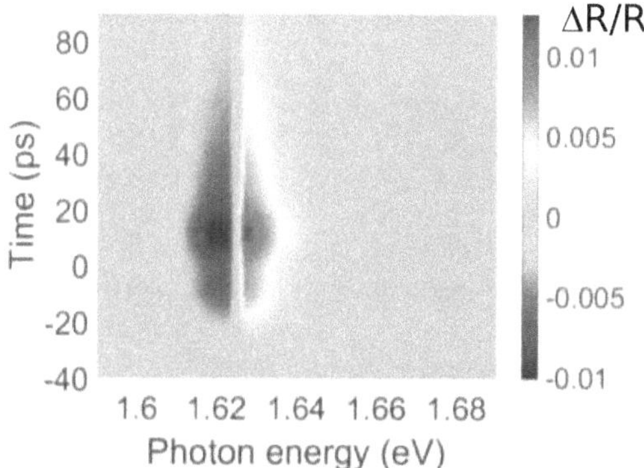

Figure 4.9.: Dynamics of the trion redshift. The experimental data are fitted with equation (4.9) using only the carrier temperature as a fitting parameter.

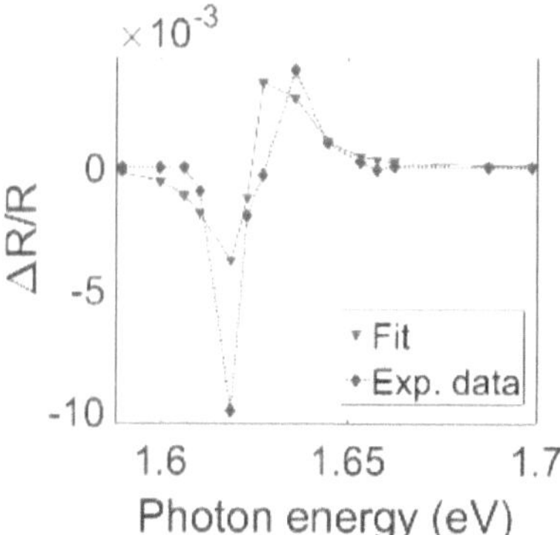

Figure 4.10.: Comparison between the experimental and calculated differential spectra at $T = 70$ ps.

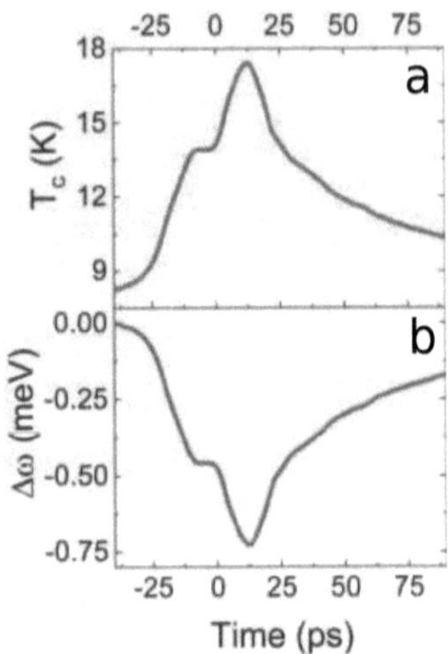

Figure 4.11.: (a) Free carrier temperature as a function of delay time. The electron cooling has a decay constant of 60 ps. (b) Trion energy shift induced by the heating of the free carrier gas.

From this procedure, it is possible to estimate the energy redshift and the carrier temperature in time (figure 4.11). The value of $T_0 = 8$ K was chosen in such a way to minimize the deviation of the fit from the data, but it was fixed at any time.

The electron cooling takes place with a bi-exponential decay. The most straightforward interpretation is that the fast component is due to the fast electron-electron scattering, while the slower component is due to the interaction with phonons and the heat transfer to the lattice.

From the change in the carrier temperature, it is possible to estimate the absorption of the infrared radiation by the monolayer. The heat capacity is roughly $c_v = \frac{3}{2} n_e k_B$. The FEL energy per pulse is $E_{pulse} = P_{FEL}/RR = 10^{-8}$ J where RR is the repetition rate of the FEL, and P is the average power. Considering also that only part of the FEL radiation hits the monolayer, the fraction of FEL energy η_{abs} transferred to free electron system, i.e. is roughly the absorption ($\eta_{abs} \backsim \alpha$), is estimated as:

$$\eta_{abs} \simeq \frac{\Delta T_c c_v V_{mono}}{E_{pulse}\left(\frac{d_{mono}}{d_{FEL}}\right)^2} \simeq 3 \cdot 10^{-3}. \tag{4.10}$$

where d_{mono} and d_{FEL} are the flake size and the FEL spot size, respectively.

This simple estimation gives a reasonable value. As counter-check, the absorption is calculated with a Drude model using the same parameters. The electron scattering time is assumed to be $\tau = 10^{-12}$ s. The absorption coefficient is $\alpha = \frac{ne^2}{\epsilon_0 c m_e \omega^2 \tau} = 1.5 \cdot 10^6$ m^{-1} and the absorption is $a = 1 - e^{-\alpha d} \simeq \alpha d = 1.3 \cdot 10^{-3}$, where $d = 0.9$ nm is the thickness of the monolayer. The two values of absorption are in good agreement, suggesting that the analysis of the trion redshift in terms of carrier temperature is a correct interpretation.

The experiment is a direct measurement of the free electron cooling. The cooling decreases exponentially with a decay constant of $\tau = 70 \pm 10$ ps. Furthermore, the redshift of the trion resonance induced by free carrier absorption is observed and studied for the first time in any semiconductor system. A quantitative description of the phenomenon is given.

4.3. Other FEL experiments on TMD crystals

In this section, two experiments performed using the infrared FEL on TMD monolayer will be presented. The first is the investigation of the intraexcitonic dynamics and the second is an experiment that attempts to detect surface phonon polariton via nanospectroscopy. The two experiments did not produce the expected outcome and the reasons will be discussed.

4.3.1. Intraexcitonic transitions in MoSe$_2$ monolayer

TMD monolayers are characterized by large exciton binding energy and large exciton oscillator strength. These two properties are ideal to study the excited states of an exciton. In fact, in a two-particle picture, an exciton can be represented with a hydrogenic model and the sequence of the excited states comes out naturally (1s, 2p, 2s,...), as it was discussed in the introduction (Sec. 1.1.2).

Not all the intraexcitonic transitions are dipole allowed but only the ones with $\Delta n = 1$ and $\Delta l = 1$, in perfect analogy with the electronic transitions in a hydrogen atom.

A substantial difference with atomic physics is the energy scale. Intraexcitonic transitions usually are in the far- and mid-infrared range. For WSe_2 monolayer for example, the 1s-to-2p transition is at 140 meV.

A couple of experiments investigating the intraexcitonic physics were carried out using the infrared free-electron laser. The first aims to explore the dynamics of the intraexcitonic transitions looking at the time-resolved PL emission. More precisely, after the excitation of an exciton population with visible pulses, the time-resolved PL emission was monitored. By adding the infrared FEL radiation tuned at the intraexcitonic transition, the population of the 1s-exciton state should be modified by the excitation of excitons in the excited 2p-state.

The second experiment aims to measure the Autler-Townes splitting induced by the resonant infrared field. The FEL wavelength was tuned in resonance with the 1s-2p exciton transition. A broadband near-infrared laser was used to probe the reflection spectra of the monolayer as a function of delay time. The Autler-Townes splitting was expected only at the time overlap of the two pulses.

In both experiments a very weak influence of the FEL radiation was observed. The exciton turned out to be very stable under infrared radiation.

Recently, Yong et al. measured the Autler-Townes splitting in a $MoSe_2$ monolayer using the same experimental approach [151]. The main experimental difference is that they use a low repetition rate table-top laser to provide an infrared source by difference frequency generation. The first data point they show is obtained using 270 kV/cm infrared field strength in vacuum and it shows a quite weak signal. The estimated optical field strength in vacuum used in the present experiments with the FEL is between 1 and 10 kV/cm. This makes the experiment unfeasible with FEL radiation.

The optical field strength needed to dress the exciton in TMD monolayers is considerably higher in comparison with other semiconductor systems, e.g. in GaAs quantum wells [158,159]. The reason of this difference is the exciton binding energy that is around one order of magnitude larger in the case of 2D TMDs. Furthermore, the absorption of a TMD monolayer in the infrared is considerably small because of the extremely small thickness of the material.

4.3.2. Surface-phonon polariton in TMD semiconductors

Surface phonon polariton (SPhP) are collective excitations in solid state materials originating from the coupling of optical phonon and photons. If the energy and the momentum of an incident photon and a phonon resonance match to each other in a regime of strong coupling or strong field, a quantum interaction between these two resonances is observed. The interaction leads to a splitting of the phonon resonance and to a non-trivial energy dispersion, following an anti-crossing behavior of the photon and phonon energy dispersion functions.

In general, a polariton can be thought as dressing of a bosonic resonance with a photon. Exciton, plasmon, or intersubband polaritons have also been observed experimentally [160, 161].

Recently, Basov and co-workers were able to perform a real-space imaging of propagating phonon polariton in thin-layer hBN via infrared nanoscopy [36]. The idea is to measure the interference between standing waves of phonon polaritons in an asymmetric geometry. The SPhPs are created under a metallic AFM tip where the field is concentrated. If the tip is close to the edge of the sample, an interference between the SPhPs launched under the tip and the SPhPs reflected by the edge is observed. A SPhP has a specific energy dispersion and the tip allows the relaxation of the momentum conservation so that phonon polaritons with various momentum can be created.

An interesting and fundamental feature of SPhPs is that they can confine the electromagnetic field overcoming the diffraction limit thanks to the strong phonon-photon coupling.

Using the same approach, an experiment of nano-imaging of SPhPs was performed on MoS_2 flakes at FELBE. Even though phonon resonances have been observed in the near-field (Figure 4.12b), no evidence of propagating phonon polaritons was observed.

The reason why no SPhPs was observed is likely the small photon-phonon coupling in TMD crystals. Limited by the small signal-to-noise ratio of the experimental setup, no significant results could be reported.

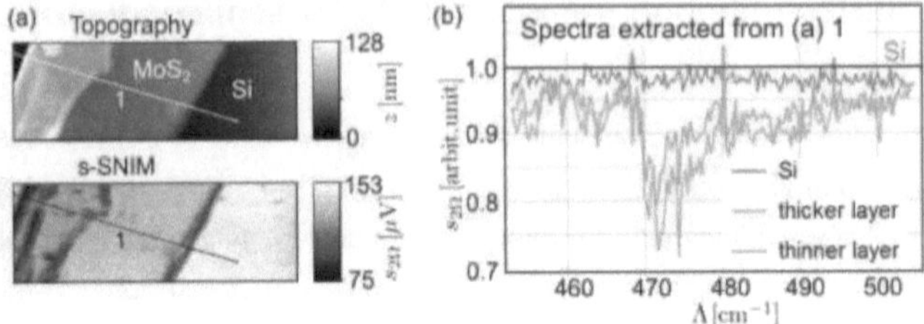

Figure 4.12.: (a) AFM topography and near-field infrared scattering at the phonon resonance in MoS$_2$ crystal ($\Lambda_{FEL} = 472\text{cm}^{-2}$). The near-field contrast depends on the thickness of the flake. (b) Near-field spectra of the A$_{2u}$ phonon resonance. The spectrum was obtained tuning point by point the FEL wavelength and measuring the near-field scattering signal along the line in figure (a).

4.4. Conclusion and outlook

Different experiments on TMD monolayers using the infrared FEL were presented in this chapter. In general, it was observed that the coupling of a bare TMD monolayer to the infrared radiation is rather weak and, most of the time, the experiments did not produce the expected results. Still, the use of two-dimensional materials for infrared applications is an interesting idea that is worth investigating further. In particular, with the use of photonic crystals, plasmonic devices, or other metamaterials the light-matter interaction can be strongly enhanced. No work is still done in this direction for the mid- and far-infrared spectral regions.

Moreover, the possibility offered by van der Waals materials to build heterostructures is a very promising avenue to engineer 2D semiconductor structures and to make infrared applications with 2D materials possible.

Nevertheless, interesting results have been obtained looking at the trion resonance. Indeed, TMD monolayers have a very large exciton and trion binding energy. As observed in different experiments, because of the very high exciton binding energy, it is difficult to affect the exciton resonance with infrared radiation. Differently, the trion binding energy is around one order of magnitude lower and infrared experiments that investigate the trion resonance are much more feasible.

A redshift of the trion resonance induced by THz radiation was observed. The effect was quantitatively explained in terms of free carrier absorption. This kind of experiment was performed for the first time on any semiconductor system and gives direct information on the free carrier absorption and subsequent cooling. Devices that measure the energy shift of the trion resonance in order to measure the incident infrared power are thinkable. A collaboration with Prof. Andreas Knorr's group of TU Berlin was established in order to get a deeper theoretical understanding of the phenomenon. A manuscript on this topic is in preparation.

5. Photoluminescence in few-layer InSe

In this chapter, the research carried out on few-layer InSe is presented. After a short introduction regarding the motivation of the work with respect to the recent research on InSe, the photoluminescence studies performed on this material are presented. The main results are:

- The fabrication of air-stable and defect-free material through encapsulation in hBN

- The measurement of temperature-dependent PL and the explanation of the PL lineshape considering exciton and free electron-hole radiative contributions.

- The measurement of time-resolved PL at low temperatures showing that the PL decay has a bi-exponential decay with time constants $\tau_1 \sim 8$ ns and $\tau_2 \sim 100$ ns.

- The observation that the PL dynamics gets slower when the number of layers is decreased because of the direct-to-indirect band-gap crossover.

- The PL decay becomes mono-exponential at high temperatures and the decay constant decreases as a consequence of more efficient non-radiative recombinations.

5.1. Introduction

As discussed in Chapter 1.2.1, InSe got increasing attention by the scientific community in the recent years. Several proof-of-principle technological devices were already tested with a good outcome as for instance field effect transistors [54,55], optical detectors [51], and bendable opto-electronics [52]. In particular, InSe showed high carrier mobility of the order of 10^3 cm^2/(Vs) due to the small electron effective mass of $m_e^* = 0.14\ m_e$ [48].

Nevertheless, InSe shows a relatively fast degradation and contamination in atmospheric conditions [162–164]. An efficient solution to this problem is to encapsulate the InSe flake in hBN. Encapsulation in hBN protects the InSe flake from oxidation but also improves the crystal quality, resulting in better optical and electrical properties. The effects of encapsulation of InSe in hBN are presented in the next section. The content of the next section is part of the work published here [35]. These results provide important steps towards the fabrication of opto-electronic devices with higher performances and good stability.

Moreover, InSe showed very interesting optical properties. One of the most striking feature is the direct-to-indirect band-gap crossover driven by the thickness of the crystal [46]. In particular, by looking at photoluminescence emission or photo-electron spectroscopy, the optical properties of few-layer InSe have been extensively discussed. However, investigations on the effects of the band-gap crossover on the dynamics were still missing. For this reason, a systematic investigation of the photoluminescence emission properties of few-layer InSe was carried out and is presented later in this chapter. The results presented in Sections 5.3 and 5.4 are partly published here [165].

5.2. Long-term stability via hBN encapsulation

The samples presented in this chapter were fabricated via mechanical exfoliation and deterministic transfer as discussed in Section 2.1. The InSe crystal was provided by Prof. Patanè's group (University of Nottingham, UK) and was grown by the Bridgman method. The development of the growth technique for InSe was done by Dr. Z. Kovalyuk and Dr. Z. Kudrynskyi. The experimental techniques are micro-PL and time-resolved PL and the details of the setups are discussed in Section 2.2.1 and Section 2.2.2.

In order to demonstrate the effects of hBN-encapsulation in few-layer InSe, the photoluminescence emission of encapsulated and unencapsulated samples is

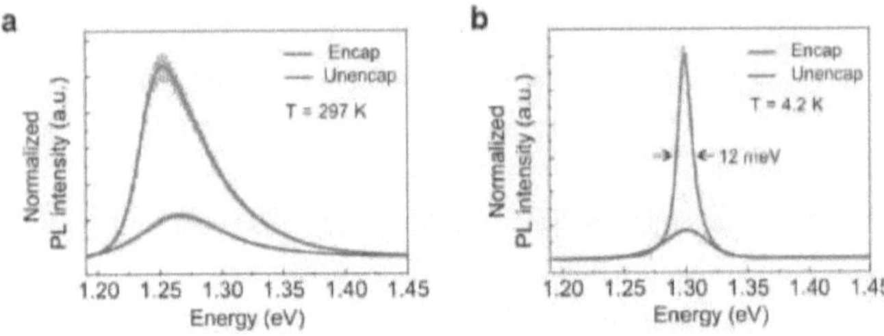

Figure 5.1.: (a) Room temperature and (b) low temperature PL of encapsulated and unencapsulated InSe flake. The PL yield of encapsulated flakes is more than five times higher. Image taken from [35].

compared. Using the micro-PL setup and, therefore, having micrometer spatial resolution, the measurements were performed on the same InSe flake that was partially encapsulated in hBN and partially unencapsulated. In this way, the uncertainty related to the sample fabrication did not play a role in the comparison.

In Figures 5.1(a) and (b), PL spectra of encapsulated and unencapsulated samples are shown at room and low temperature, respectively. The measurements were performed few hours after the sample fabrication. The encapsulated InSe shows much higher PL yield with respect to the unencapsulated sample (five times higher), that means less non-radiative recombination. This is an indication that the oxidation of the InSe flake already plays a significant role over a span of hours.

The different dielectric environment introduced by the hBN layers can also modify the PL intensity. However, the hBN layers were $10 - 30$ nm thick. The excitation wavelength and the PL wavelength are much longer than the thickness of the hBN layers, therefore the effects of interference are not very prominent.

More evidently, at low temperature the FWHM of the PL emission is reduced from 35 meV for the unencapsulated area to 12 meV for the encapsulated area. This quantity is a direct measurement of the disorder potential in the crystal. In fact, the inhomogeneous broadening of the PL line is due to all the imperfections in the crystal lattice. More specifically, the disorder potential is induced by the substrate, i.e. roughness and charge transfer [166], and by oxidation and adsoprtion of gas molecules [167]. These effects are strongly reduced by the encapsulation in hBN.

Afterwards, the long-term stability of the samples in ambient condition was

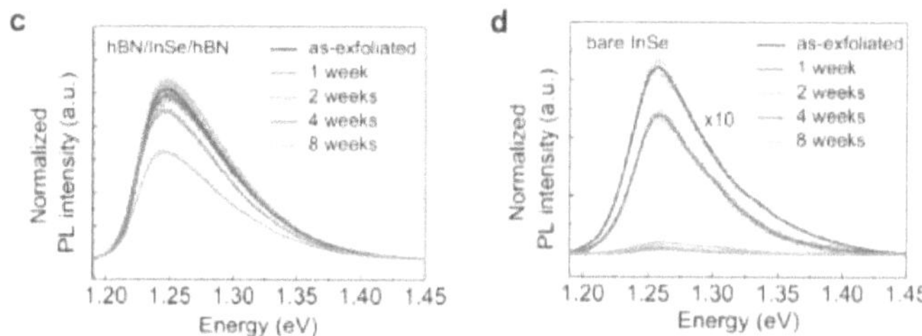

Figure 5.2.: (a) and (b) room temperature PL of encapsulated and unencapsulated InSe flake respectively over a span of eight weeks. Image taken from [35].

monitored by room temperature PL. In figure 5.2, the PL spectra of encapsulated and unencapsulated samples are shown. The unencapsulated shows an abrupt degradation in the PL emission while the encapsulated samples maintain a robust PL emission. This demonstrates the capability of encapsulation in hBN to prevent the aging of thin-layer InSe.

5.3. Steady-state photoluminescence

After the successful fabrication of hBN-encapsulated InSe samples, a detailed investigation of their PL emission properties were carried out.

5.3.1. Temperature dependence of PL

Figures 5.3(a) and 5.3(b) show temperature-dependent PL spectra for 24- and 9-layer InSe crystals, respectively. The energy position of the PL bands of the two samples at 4 K is in agreement with previous reports on thin layers of InSe [168]. At $T = 4$ K both samples show a single broad PL emission band mostly due to defect-assisted radiative recombination. As noted in [40], the large broadening of the PL lines is due to the low electron mass that makes the emission very sensitive to any surface effect and to any disorder potential. The broadening of the PL lines increases while decreasing the number of layers, as becomes clear from the broad emission of the 9-layer sample.

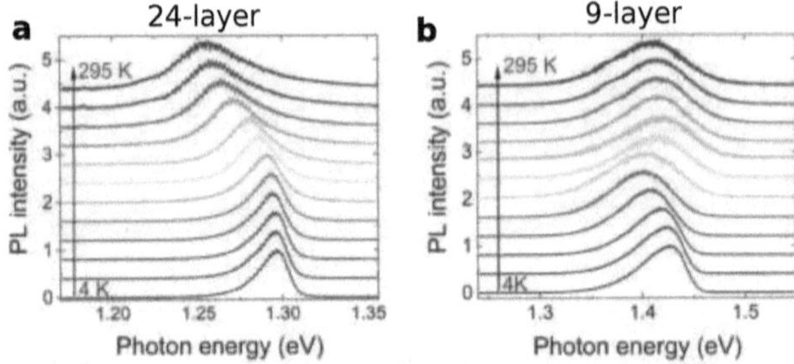

Figure 5.3.: (a) and (b) Temperature dependence of the PL spectra of a 24-layer and 9-layer InSe samples, respectively. The excitation power is 10 μW. For the 24-layer thick sample a redshift of the PL emission is observed due to the reduction of the band-gap energy. The 9-layer sample shows an s-shape due to the large contribution of the defect-assisted recombination.

While increasing the temperature, the 24-layer sample shows a monotonic redshift that is a consequence of the reduction of the band-gap energy due to the lattice expansion and to the interaction with phonons. The dependence of the band-gap energy on temperature is analyzed in Appendix A.5 by using the empirical O'Donnell model. The temperature dependence of the PL energy position of the 9-layer flake shows an s-shape. This is a consequence of the defect-state emission that dominates the PL at low temperature, as discussed in previous studies and observed in other semiconductor systems [168, 169]. The PL energy position of the 9-layer sample shows an overall blueshift in comparison with the thicker InSe sample. This is due to quantum confinement and reduction of the dielectric screening [40, 170].

5.3.2. Lineshape analysis

In order to get more information on the PL emission mechanisms, a model for the PL lineshape was developed starting from the model proposed by Katahara and Hillhouse [128]. This model is a generalized version of the van Roosbroeck-Shockley equation that connects absorption and interband PL in semiconductors [171, 172]. Knowing the absorption of the material and the temperature of the system, it is possible to calculate the PL emission:

$$I_{PL}(E) \propto \frac{E^2 a(E)}{\exp\left(\frac{E-\Delta\mu}{kT_{PL}}\right) - 1} \cdot \left(1 - \frac{2}{\exp\left(\frac{E-\Delta\mu}{2kT_{PL}}\right) + 1}\right) \tag{5.1}$$

where the first part is the connection between Planck's law and the absorption and the second part in parentheses is a small correction that takes into account the occupation of the bands (Pauli blocking). $a(E) = a_B(E) + p\, a_X(E)$ is the total absorption given by a linear combination of band-to-band and exciton contribution, $\Delta\mu$ is the quasi-Fermi energy, and T_{PL} is the effective photoluminescence temperature. The quasi-Fermi energy is an effective Fermi energy introduced by Lasher *et al.* in order to consider the occupation of the conduction and valence band [172].

Neglecting the correction due to the occupation of the bands, the part of Equation (5.1) containing the temperature can be simplified assuming a Boltzmann distribution.

The key element of the model is the absorption. For the exciton absorption, a Gaussian shape was assumed:

$$\alpha_X(E) \propto \frac{1}{\sqrt{\pi\sigma^2}} e^{-\left(\frac{E-E_X}{\sigma}\right)^2} \tag{5.2}$$

where E_X is the exciton energy, and σ is the broadening of the exciton resonance. The exciton lineshape derived by the semiconductor Bloch equations (1.2) is Lorentzian. However, often a Gaussian-distributed inhomogeneous broadening is present and a Gaussian lineshape is the correct shape for an exciton resonance even at zero Kelvin [173].

For the electron-hole absorption, the convolution of the free electron-hole 3D density of states ($D(E) \propto \sqrt{E - E_g}$) and an Urbach tail ($U(E) = E_u \exp\left(-\frac{E}{E_u}\right)\theta(E)$) was used. The following is the result of the convolution and the expression used in the model:

$$\alpha_B(E) \propto E_u^{\frac{5}{2}} \begin{cases} e^{\frac{E-E_g}{E_u}} & : E < E_g \\ \sqrt{\frac{4(E-E_g)}{\pi E_u}} + \mathrm{erfc}\left(\sqrt{\frac{E-E_g}{E_u}}\right)e^{\frac{E-E_g}{E_u}} & : E \geq E_g \end{cases} \tag{5.3}$$

where erfc() is the complementary error function. For energy below the band

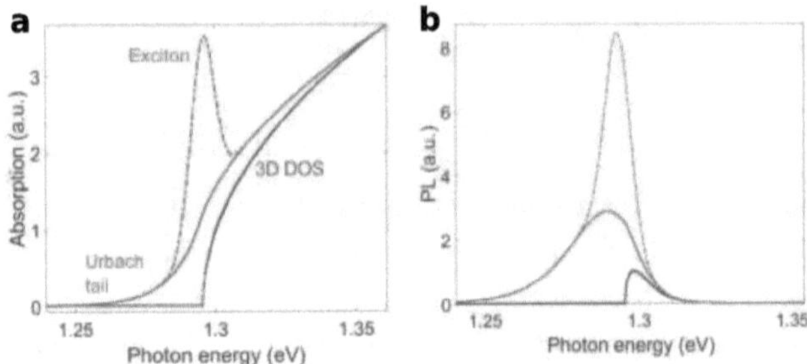

Figure 5.4.: (a) Absorption related to the three dimensional DOS, Urbach tail and exciton. (b) PL calculated with Equation (5.1). The PL emission is given by the thermal occupation of the states shown in absorption.

gap, the function is an exponential rise (Urbach) and above is the usual 3D DOS. The last term assures that the function belongs to the class of C_1 functions (the function is differentiable).

Considering also the small thickness of the sample, a linear relation between absorption and absorptivity $a(E) = \alpha(E)d$ is assumed, where d is the thickness of the sample. So finally, by plugging $a(E) = a_B(E) + p\, a_X(E)$ into Equation (3.1), an analyitical expression for the PL lineshape is obtained.

To get a visual understanding of the model, Figure 5.4 shows the different components to absorption and the resulting PL spectra.

Hence, the PL spectra at each temperature in Figure 5.3a are fitted with Equation (5.1). The fitting function can reproduce the PL data for each temperature with excellent accuracy as shown for example in Figure 5.5(a) for the spectrum at $T = 4$ K (red curve). Moreover, the fit obtained without the exciton contribution is also shown for comparison. Figure 5.5(b) shows the free electron-hole and exciton contributions separately.

The exciton contribution to the PL emission is a fundamental correction that has to be taken into account. However, at high temperature, the simple electron-hole function can reproduce very well the experimental data.

Figure 5.6 shows the absorption spectra as obtained from the fit for each temperature. At low temperatures, the exciton resonance is clearly observable in the absorption spectra and it smears out at higher temperature as expected. In

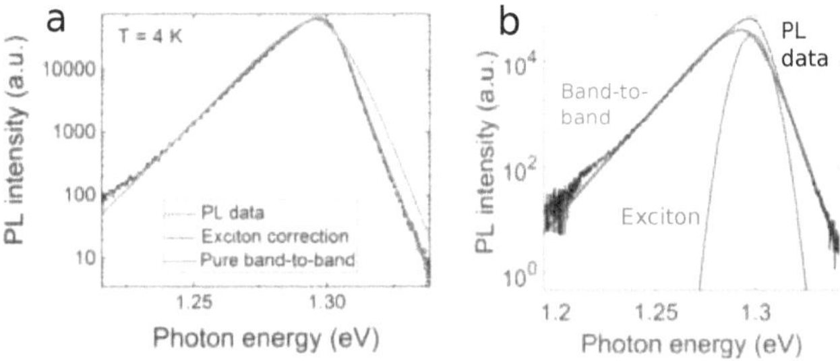

Figure 5.5.: (a) PL emission at 4 K with fit with and without the exciton correction. (b) Decomposition of the free electron-hole and exciton contributions as obtained by the model described in the text.

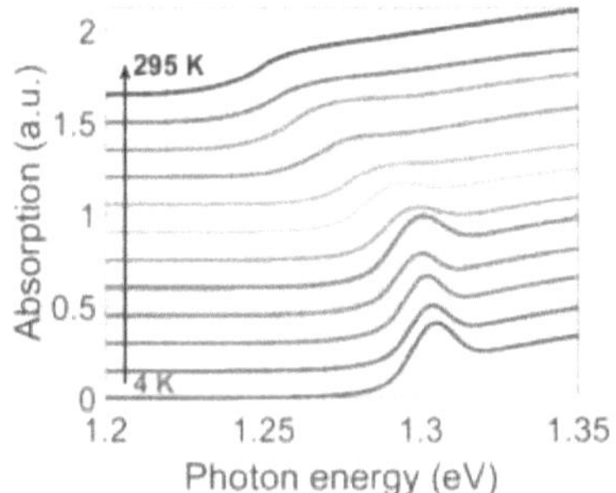

Figure 5.6.: Absorption spectra as extracted from the fitting of the PL spectra.

fact, the exciton binding energy is around 14 meV which corresponds to about 160 K [174–176].

Photoluminescence temperature

An interesting detail of the model is the introduction of an effective PL temperature in order to reproduce accurately the PL spectra at low temperatures. In general, for direct band-gap semiconductors, the high-energy side of a PL peak gives a direct information on the electron-hole temperature. In fact, the slope of the high-energy side of the peak is given by the tail of the thermalized electron-hole distribution in the bands. At high temperatures, the main mechanism of

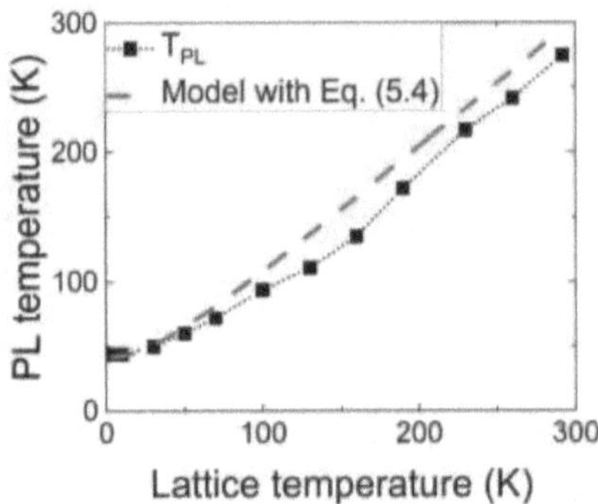

Figure 5.7.: Effective PL temperature obtained from the model with respect to the actual lattice temperature.

broadening of the PL line is thermal and a long tail on the high energy side is usually observed, as for instance in Figure 5.3(a). Differently, at low temperatures, the inhomogeneous broadening is dominant because the thermal distribution is limited on the high energy side by a sharp Fermi edge. However, using the temperature as a parameter, it is possible to reproduce the inhomogeneous broadening at low temperatures.

Figure 5.7 shows the PL temperature with respect to the lattice temperature. The PL temperature does not approach zero Kelvin but it saturates around a certain value greater than zero.

A model for the PL temperature is:

$$T_{PL} = \sqrt{T^2 + T_0^2},\qquad(5.4)$$

in analogy with the model proposed by Marianer et al. for disordered semiconductors [96, 131]. This formula can reproduce the data with excellent agreement.

The idea of this procedure is to incorporate the disorder, that shows up as inhomogeneous broadening of the PL line, in the effective PL temperature. Therefore, $T_0 = 44$ K or $k_B T_0 = 4$ meV is a measurement of the disorder in the material and it can be intuitively interpreted as the standard deviation of the disorder potential in the sample due to lattice defects and sample inhomogeneity.

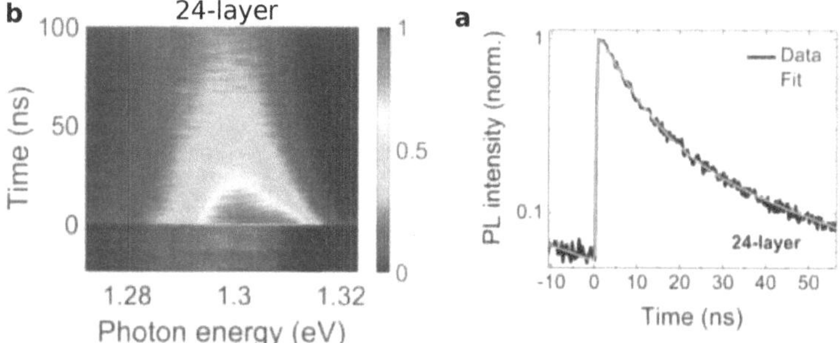

Figure 5.8.: (a) 2D false-color maps of normalized PL intensity as a function of time and photon energy at $T = 4$ K and 10 μJ cm^{-2} excitation fluence. The sample is 24-layer thick (b) PL decays integrated in photon energy. The decay is bi-exponential and has time constants $\tau_1 = 7.7 \pm 0.2$ ns and $\tau_1 = 49 \pm 6$ ns.

This treatment has a general validity for semiconductors with direct band gap.

It should be noted that a non-vanishing T_0 could be also due to higher electron-hole temperature induced by laser excitation. However, in this case, the effect is negligible because no dependence of the parameter T_0 on excitation power was observed (not shown).

5.4. Time-resolved photoluminescence

Figure 5.8(a) shows the PL intensity as a function of time and photon energy for a 24-layer InSe crystal. The PL emission shows a spectrally dependent bi-exponential decay. In order to get the time constants, the PL emission was integrated over the photon energy and the data were fitted with a bi-exponential decay convoluted with the instrument response function. The energy-integrated PL decay is shown in Figure 5.8(b). The fast and slow PL components have lifetime of $\tau_1 = 7.7 \pm 0.2$ ns and $\tau_1 = 49 \pm 6$ ns, respectively. In order to visualize the two PL components independently, 2D false-color plots of the extracted fast and slow decays are shown in Appendix A.6.

The PL decay in Figure 5.8 has a small contribution visible at time less than zero. This is due to the excitation from the previous pulse. In fact, the repetition

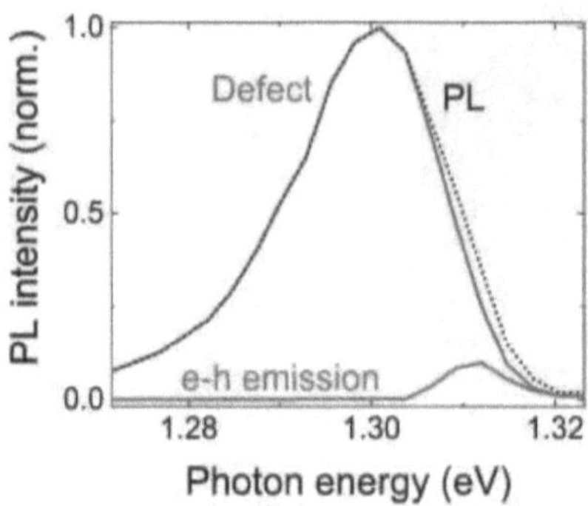

Figure 5.9.: Spectra of the extracted fast (electron-hole) and slow (defect) components for the 24-layer thick sample.

rate of laser is 13 MHz, i.e. one pulse every 78 ns.

The slow PL decay is a fingerprint of a defect-assisted electron-hole recombination. Therefore, the slow component is associated with this radiative channel. The fast component is associated with the direct-band-gap electron-hole recombination.

By integrating in time the fast and the slow components as obtained from the fit, it is possible to separate the spectra of the two contributions (shown in Figure 5.9). The central emission energy of the fast component is 1.312 ± 0.001 eV with full width at half maximum (FWHM) of 6.8 ± 0.5 meV, while the slow component is centered at 1.300 ± 0.001 eV with FWHM of 15 ± 1 meV. The slow component lies at lower energy and with a broader spectrum, as expected for a defect-assisted emission. The two components were not spectrally distinguishable in steady-state PL.

Now, the layer dependence of the time-resolved PL is investigated. Figure 5.10(a) shows the PL data for a 9-layer thick InSe crystal measured under the same condition as the 24-layer thick sample. Qualitatively, it is evident that the PL decay takes place on a longer timescale. In addition, the weights of the two PL components change significantly. The defect-assisted recombination appears to be more dominant with respect to the electron-hole recombination for thinner samples, as expected from the direct-to-indirect band-gap crossover driven by the

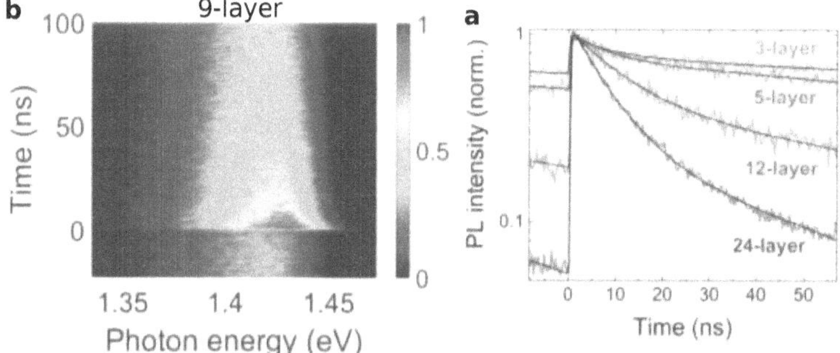

Figure 5.10.: (a) 2D false-color maps of normalized PL intensity as a function of time and photon energy at $T = 4$ K and 10 μJ cm^{-2} excitation fluence. The sample is 9-layer thick. (b) PL decays integrated in photon energy of samples with different thicknesses. The long-living component of the PL decay becomes more prominent for thinner flakes. This is a signature of the direct-to-incdirect band-gap crossover.

sample thickness [40]. Conversely, electrons and holes in the 24-layer sample can recombine easily because of the direct band gap, while in thin samples they need the assistance of a scattering process.

To further corroborate this observation and to perform a quantitative analysis, a series of encapsulated InSe crystals with different thicknesses were fabricated and measured. Figure 5.10(b) shows the PL decays integrated in photon energy for different layer thicknesses.

The PL decays were fitted with a bi-exponential decay function and the fast and slow lifetimes (τ_1 and τ_2) were extracted for each sample. The lifetimes are shown in Figure 5.11(a). The fast component, i.e. electron-hole recombination, does not show any clear layer-dependence. On the other hand, the lifetime of the slow component increases upon decreasing the flake thickness. The dynamics of the electron-hole recombination is limited at low temperature mainly by the population transfer to lower energy states, i.e. bound states below the band gap. This could explain why no layer dependence was observed for the fast component of the decay.

The ratio $\frac{I(\tau_1)}{I(\tau_2)}$ is shown in Figure 5.11(b) as function of layer thickness. The

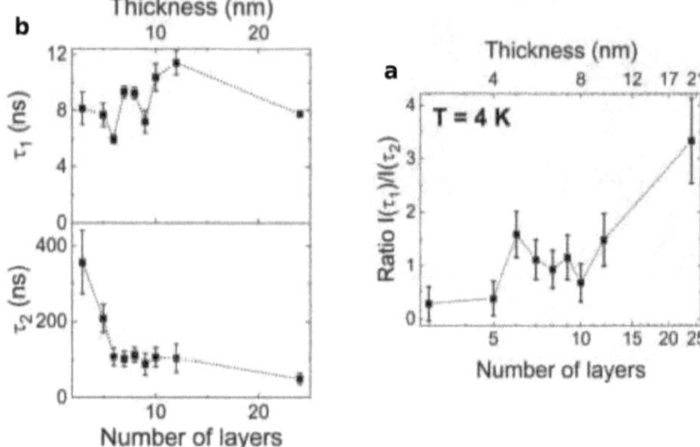

Figure 5.11.: (a) Lifetimes of the fast and slow components of the PL decay as a function of sample thickness. (b) Dependence of the ratio between PL contributions associated with the fast and the slow component on the number of layers.

fast component becomes more and more dominant with increasing the number of layers. The increase of the ratio takes place between five and ten layers, i.e. in good agreement with the number of layers which the direct-to-indirect band-gap crossover is expected at [46]. The error bars were obtained from the standard errors of the least-square fitting procedure and the propagation of errors.

Finally, the temperature dependence of the PL dynamics is studied. Figure 5.12(a) shows the electron-hole PL decays as a function of temperature. The PL lifetime decreases monotonically with increasing the temperature due to higher non-radiative recombination, i.e. by phonon scattering. Remarkably, the PL decays at temperatures higher than 70 K show a single exponential decay. This is because the defect states responsible for the slow component are not stable anymore and the only radiative channel is the fast electron-hole recombination.

Figure 5.12(b) shows the extracted PL lifetime as a function of temperature. An empirical model is used in order to fit the data. Radiative and non-radiative decays are considered. Assuming an exponential decay, the lifetime is inversely

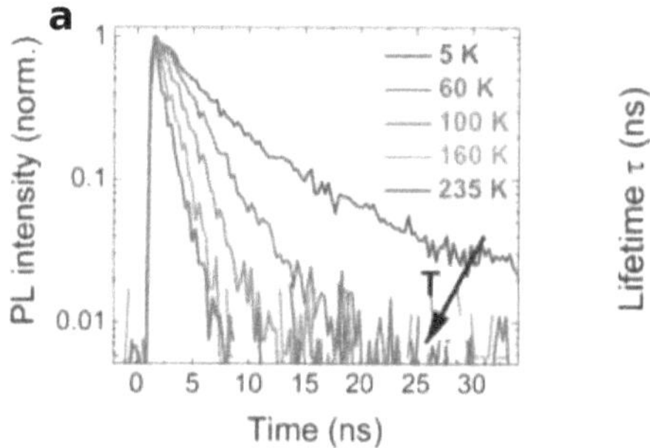

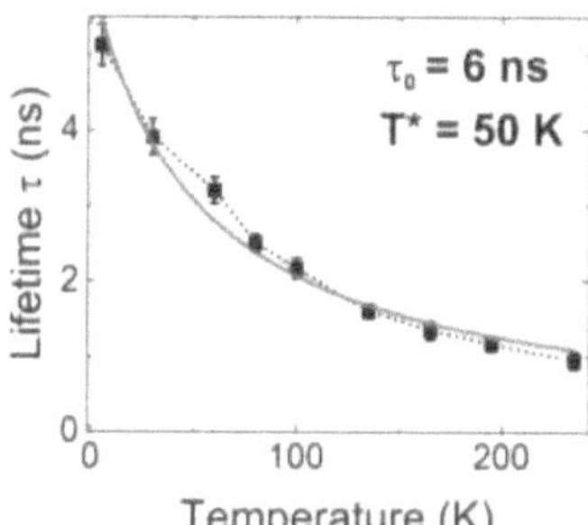

Figure 5.12.: (a) PL decays as a function of temperature. The PL decay gets faster at higher temperature due to the more efficient non-radiative scattering. The sample is 24-layer thick and fully encapsulated in hBN. (b) PL lifetime τ_1 as a function of temperature. The red curve is obtained from equation 5.5.

proportional to the sum of the radiative and non-radiative decay rates:

$$\tau = \frac{1}{\beta_r + \beta_{nr}} = \frac{1}{\beta_0 + \alpha_{nr}T} \tag{5.5}$$

where β_r and β_{nr} are, respectively, the radiative and non-radiative decay rates, β_0 is the decay coefficient at zero Kelvin, and a linear approximation was done to model the dependence of the non-radiative scattering on temperature.

This simple approximation shows a good agreement with the experimental data and gives two quantitative pieces of information: 1) the lifetime at zero Kelvin ($\frac{1}{\beta_0} = \tau_0 = 6$ ns) and 2) the temperature T^* at which the non-radiative scattering overcomes the radiative scattering. The latter is: $T^* = \frac{\beta_0}{\alpha_{nr}} = 50$ K.

If a square root temperature dependence were used in Equation (5.5) instead of a linear one, the model would look like the Shockley-Read-Hall model for non-radiative recombination [177,178]. However, this gives a worse agreement with the experimental data and no significant parameters can be extracted. Furthermore, according to the data, the best-fitting exponent for the temperature would be $1.34 \pm 0.06 \approx \frac{4}{3}$ instead of 1, but no physical explanation could be found.

5.5. Conclusions and outlook

III-VI semiconductors are a subset of van der Waals semiconductors which have different lattice structure and opto-electronic properties in comparison with TMD monolayers, making the family of 2D semiconductors more diverse and valuable. During the course of the PhD, research was conducted on different materials which belong to this class. In this chapter, the research performed on InSe was presented.

Firstly, the fabrication of air-stable samples via hBN-encapsulation was demonstrated. Afterwards, the PL properties of thin-layer InSe were investigated and the involved microscopic physical mechanisms were discussed.

By measuring the temperature dependence of the PL emission, the role of excitonic effects was discussed with respect to the free electron-hole recombination. The exciton features are visible in the PL spectra up to 160 K= 14 meV. Remarkably, the model presented for this analysis has a general validity for direct band-gap semiconductors.

The analysis of the time-resolved PL signals allows us to disentangle the contributions from direct band-gap electron-hole recombination ($\tau_1 \sim 8$ ns) and from defect-assisted recombination ($\tau_2 \sim 100$ ns). These contributions are not spectrally distinguishable without resolving the dynamics. The defect-assisted contribution becomes increasingly important as the number of layers is decreased. Remarkably, the electron-hole PL lifetime basically is independent of the number of layers, while the lifetime of the defect-assisted recombination increases for thinner samples. Furthermore, shorter PL lifetimes were found with increasing temperature, which is caused by more efficient non-radiative recombination.

These discoveries are important for technological applications based on few-layer InSe. The demonstration of the positive effects of hBN-encapsulation on the performances of opto-electronic devices has a direct impact on applications. Nonetheless, the knowledge of the microscopic mechanisms involved in the light-matter interaction is a vital information for the design of devices.

Research on InSe is an on-going project. In particular, an experiment of PL quenching induced by infrared free-electron laser radiation was already performed obtaining promising results. Furthermore, other interesting systems are planned to be investigated as for example InSe/TMD heterostructures.

6. TMD heterostructures

In the last part of the PhD, heterostructures of WSe_2 and $MoSe_2$ monolayers were fabricated and studied. This work is the preliminary part of an on-going project that aims to investigate the infrared response of $WSe_2/MoSe_2$ monolayer heterostructures using the FEL at FELBE.

In this chapter, the fabrication of heterostructures and some basic spectroscopic results will be presented.

6.1. Twisted angle heterostructures

The fabrication of $WSe_2/MoSe_2$ monolayer heterostructures was carried out using the PDMS stamp technique as described in Section 2.1.2. This technique provides samples with high optical quality. However, the size of the samples is on the micrometer scale and this is a limitation for infrared spectroscopy.

The two compounds have the same lattice structure. Furthermore, the mismatch between the lattice constants of the two monolayers is very small. The lattice constant of WSe_2 is 0.330 nm and of $MoSe_2$ is 0.328 nm [179]. The difference between the two lattice constants is around 0.2%. For this reason, the hybridization between the wavefunctions of the two monolayers is very efficient.

Remarkably, the fabrication method of deterministic transfer via PDMS gives an additional degree of freedom, i.e. the relative angle between the crystallographic axes of the two compounds. This is called twist angle. The twist angle determines the electronic coupling between the two compounds. Calling θ the twist angle, the maximum coupling is obtained in the case of $\theta = 0° + n \cdot 120°$ or $\theta = 60° + n \cdot 120°$, where n is an integer number [180]. The distinction between 0° and 60° is due to the fact that the K and K' valleys of WSe_2 and $MoSe_2$ monolayers are not identical because of the inversion symmetry breaking.

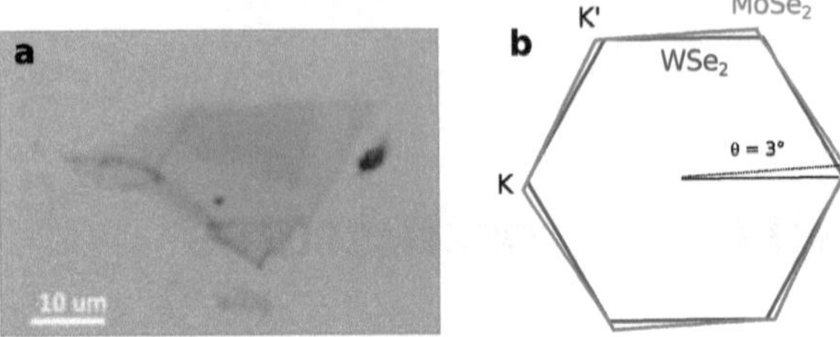

Figure 6.1.: (a) Optical image of a WSe$_2$/MoSe$_2$ heterostructure. (b) Reciprocal space alignment.

A striking effect related to the twist angle is the formation of a moiré pattern in the heterostructures. The moiré is formed for heterostructures also in the case of 0° because the different materials have different lattice constant. In the case of WSe$_2$/MoSe$_2$ heterostructure, the period of the moiré superlattice is around 100 nm at $\theta = 0$. For angles greater than zero, the size scale of the pattern becomes shorter and shorter. The angle at which the moiré period is minimum is 30°.

The moiré pattern is associated with a moiré potential and a corrugation in real space of the lattice. The moiré potential is periodic over the crystal and has a maximum energetic depth of around 100 meV, as calculated numerically [66]. It has been shown that the moiré potential is able to localize excitons [66, 181]. By varying the twist angle, it is possible to control the moiré potential and ideally it is possible to engineer the properties of the heterostructures. A natural lattice of localized quantum dots has been proposed with this scheme [182].

The twist angle does not only control the moiré potential, but also changes the hybridization of the wavefunctions of the layers via small but significant covalent bonds. One of the most spectacular results is the unconventional superconductivity observed in twisted bilayer graphene [60].

For WSe$_2$/MoSe$_2$ heterostructures, the optimum coupling angle is reached when the crystallographic axes are aligned (0° or 60°). The alignment in real space will result in alignment in the reciprocal space and the excitons in the K(K') points interact efficiently.

Using a micro-manipulator with a rotational mounting, it is possible to control experimentally the angle between the two monolayers with very good accuracy.

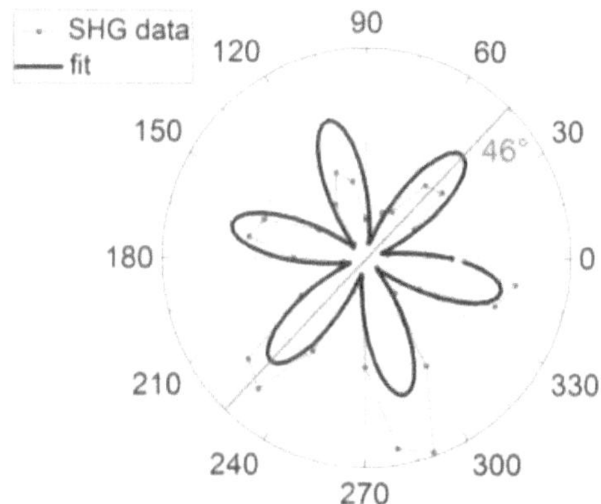

Figure 6.2.: Polarization dependent second harmonic generation from MoSe$_2$ monolayer on PDMS. Every maximum of the sixfold degenerate pattern indicates the armchair direction of the lattice.

However, the knowledge of the orientation of the crystallographic axes is required. A first rough indication of the axes orientation is given by the edges of the monolayer flakes. It is likely that the edges follow the crystallographic edges. Therefore, by aligning the edges of the two monolayers at the microscope, there are good chances to get the correct alignment of the axes. One of these cases is shown in Figure 6.1.

This idea works only partially and it cannot be used reliably. In order to know the crystallographic axes of each monolayer before stacking, polarization-dependent second harmonic generation (SHG) can be used [183, 184]. The intensity of the SHG component with polarization parallel to the excitation source has a strong dependence on the relative orientation with respect to the crystal axes. More precisely, $I_{SHG} = I_0 \cos^2(3\phi)$, where ϕ is the relative angle between polarization of the excitation source and the lattice axes. If there is a perfect match (0° or 60°) between the linear polarization of the laser and the lattice, the SHG is strongly enhanced.

Figure 6.2 shows the polarization dependence of the SHG of a MoSe$_2$ monolayer flake. A typical six-fold flower shape is observed.

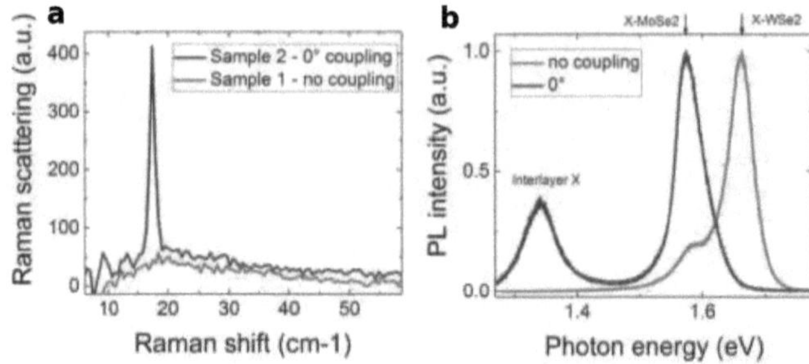

Figure 6.3.: (a) Raman spectra of two different stacking angles. When the alignment of the two monolayers is close to 0° or 60°, a low-frequency Raman peak is observed. (b) PL spectra at room temperature of the same devices.

For the fabrication of heterostructures, the polarization dependence of the SHG is measured in each monolayer after the exfoliation. The monolayers are still on PDMS. Only after the knowledge of the axes has been obtained, the monolayers are stacked on top of each other.

This method is very reliable and efficient. However, it is not able to distinguish between K and K' valley, i.e. 0° or 60° alignment. Another limitation of this method is that it can be used only for monolayers. Bilayer and bulk samples of TMD do not have a strong second order non-linear susceptibility because they have the inversion symmetry and the even harmonics are suppressed.

After the fabrication of the heterostructures, the effective coupling is tested with Raman and PL spectroscopy. Figure 6.3 shows low-frequency Raman spectra of two heterostructures. In the case of nearly 0° angle, an intense shear vibrational mode is observed at 17 cm^{-1}. The origin is the coupling between the z-orbitals of the selenium atoms in the different monolayers [180]. This is already a perfect tool to verify the quality of the fabricated heterostructure.

Furthermore, a new peak appears in PL at 1.35 eV in the case of good alignment of the two lattices (Figure 6.3). This peak is due to the formation and radiative recombination of interlayer excitons [185].

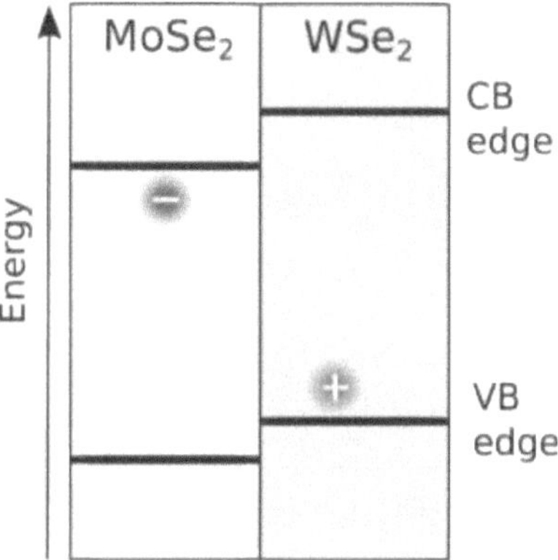

Figure 6.4.: Sketch of the band alignment of WSe$_2$ and MoSe$_2$ monolayers. Electrons and holes relax to the lowest energy states and they form interlayer excitons.

6.2. Photoluminescence of interlayer excitons

An interlayer exciton is an electron-hole pair that lies partly in one material of the heterostructure and partly in the other. In the case of MoSe$_2$/WSe$_2$ monolayer heterostructures, the photo-excited electrons and holes relax quickly into the lowest energy bands of the two materials, i.e. the electrons transfer into the MoSe$_2$ monolayer and the holes into the WSe$_2$ monolayer (sketch in Figure 6.4). Even though electrons and holes lie in the two different layers, they still interact via Coulomb force forming a bound pair, properly called interlayer exciton.

The energy position of the interlayer exciton at room temperature is 1.35 eV. This value is obtained by the energy gap from the conduction band of MoSe$_2$ to the valence band of WSe$_2$ subtracting the interlayer exciton binding energy. On top of this, the perturbation induced by the moiré potential further decreases the observed energy position.

In case of lattice misalignment, the interlayer exciton peak is hardly observable. This is because the misalignment in the real space translates into a misalignment in the reciprocal space and the interlayer excitons become indirect in momentum space.

Figure 6.5 shows the temperature dependence of the PL spectra of a device with

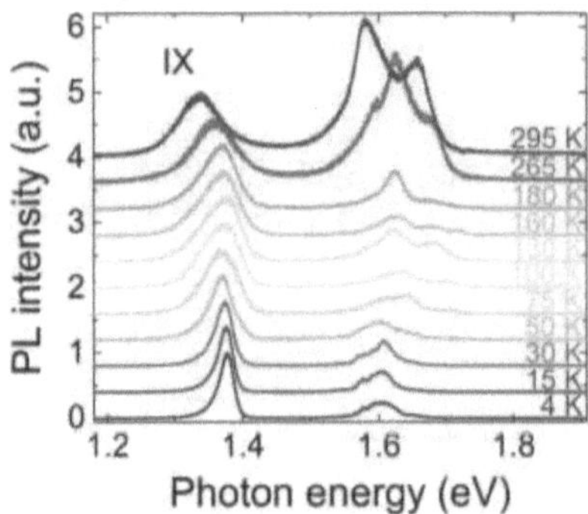

Figure 6.5.: Temperature dependence of $MoSe_2$/WSe_2 heterostructure with good axes alignment. The temperature dependence of the interlayer exciton shows an s-shape due to exciton localization of the moiré potential at low temperatures.

close to 0° alignment. At low temperatures, the interlayer exciton is the dominant feature of the spectra, while at room temperature the intralayer exciton of the individual monolayers are more prominent.

The temperature dependence of the energy position of the interlayer exciton is rather complicated. The energy shift with temperature is not monotonic suggesting the presence of a defect-assisted component at low temperature and other competitive mechanisms [186]. Furthermore, the moiré potential creates a landscape of potential wells in the heterostructures that are able to localize excitons. Hopping between potential wells can play a significant role.

The temperature dependence of the intralayer exciton is non monotonic because at low temperatures only the localized exciton emission is visible. Furthermore, trion resonances both in the $MoSe_2$ and the WSe_2 layers are also visible.

An interesting characteristic of interlayer excitons is that many-body quasiparticles as trions or biexcitons are less likely to be observed. In fact, the carriers in each layer repel each other avoiding the formation of many-particle systems, such as trions and biexcitons. For this reason, high temperature exciton condensation has recently been observed [65].

Time-resolved PL (trPL) was also performed using a streak-camera. Figure 6.6

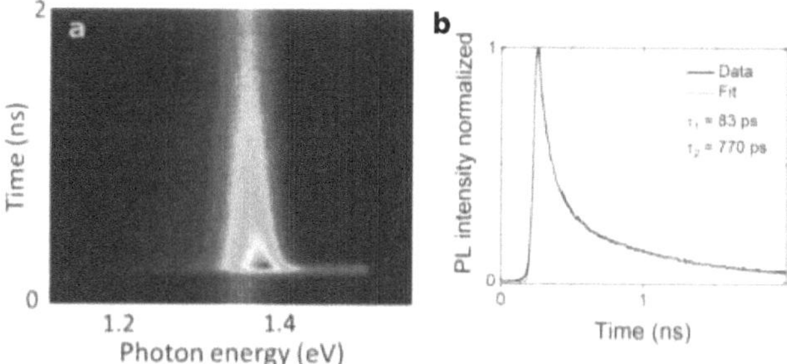

Figure 6.6.: (a) Time-resolved PL spectra at $T = 5$ K. The PL decay takes place on a time scale of nanoseconds. (b) PL decay integrated in photon energy. The decay is fitted by a bi-exponential function.

shows the trPL spectrum. At high fluences ($\phi = 30 \ \mu \text{Jcm}^{-2}$), a bi-exponential decay was observed with time constants $\tau_1 = 83$ ps and $\tau_2 = 770$ ps. The power dependence of the interlayer exciton PL decay is shown in Figure 6.7. At low fluences, a mono-exponential decay is observed. The fast component is therefore tentatively attributed to exciton-exciton scattering that happens at high exciton densities.

But in general, interlayer excitons have much longer lifetime in comparison with the intralayer counterpart. This is because the interlayer excitons are spatially indirect. Nevertheless, this feature makes them very promising for different technologies, for instance heterostructure lasers, photovoltaic devices, or valleytronic quantum computation.

The work presented in this chapter is the preliminary step for research on TMD heterostructures. With the time-resolved PL setup, a FEL experiment is planned. The FEL will be tuned at the excited states of the interlayer exciton and a quenching and recovery of PL emission is expected.

Other experiments are also in schedule as for instance the investigation of hopping of localized excitons in the moiré potential or the possible role of intersubband transition in a three-layer device.

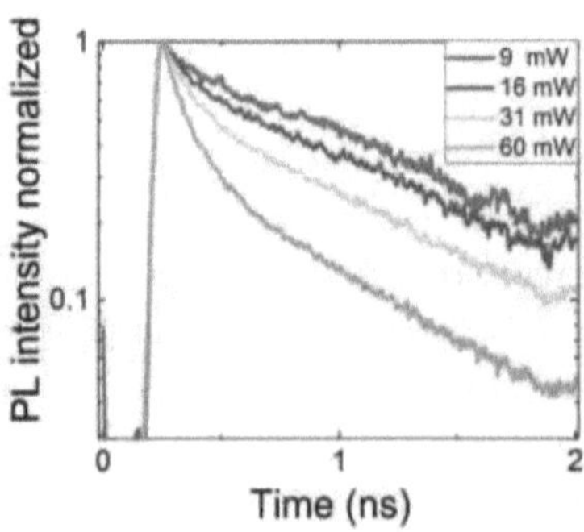

Figure 6.7.: Power dependence of the PL decay of interlayer exciton at 5 K.

7. Conclusion and outlook

This **book** has focused on the optical properties of atomically thin semicon-ductors. The first purpose of this **book** is to give a comprehensive introduction and a general overview of the most fascinating and promising properties of these materials as well as a discussion on their technological issues and limitations.

The first experimental result is related to the fabrication of monolayer samples of different kinds of transition metal dichalcogenides. This was the starting point of the research and continuous improvements in the fabrication techniques have been developed as described in Chapter 2.1.

With the in-house fabricated samples, many different optical experiments have been performed, deepening the comprehension of the physics of these materials. The role of exciton localization in the photoluminescence spectra of $MoSe_2$ mono-layer was considered and carefully studied. In particular, it has been shown that gas molecules adsorbed on the monolayer surface are able to localize excitons at low temperatures and, furthermore, that via laser irradiation it is possible to break the physisorption binding of the molecules and release the localized exci-tons. These results were reported and discussed in Chapter 3.

These last results were fundamental in order to understand the other experi-ments on this material. In particular, a THz-pump optical-probe experiment was carried out with the infrared free-electron laser: without the analysis of the local-ization of excitons and the effects of laser irradiation, the comprehension of the unexpected behavior observed in this experiment would not have been clarified.

In fact, in this pump-probe experiment a redshift of the trion resonance was ob-served as consequence of infrared non-resonant radiation. As described in detail in Chapter 4, the redshift is caused by infrared free carrier absorption. Con-versely, the free-electron population is accelerated by the infrared absorption and the kinetic energy and the momentum of the carriers is transferred to the trion population when the trion population is created by the optical pulse. The trion transition energy is reduced by the kinetic energy of the free electron.

A careful quantitative analysis of this process was conducted using the density matrix formalism. The model shows an excellent agreement with the experimental data and gives us a deeper and more solid comprehension of this phenomenon.

Besides the aforementioned experiments on TMD monolayers, research on other van der Waals semiconductors was carried out. In particular, the experiments on InSe thin flakes were reported in Chapter 5. Since InSe degrades quite quickly in air under ambient condition, the flakes were encapsulated in hBN and the optical properties of such flakes were carefully investigated. The positive effects of encapsulation were demonstrated showing that hBN-encapsulation both protects InSe from degradation and at the same time improves the optical and electrical performances.

In addition, the dynamics of the photo-excited electron-hole system in hBN-encapsulated InSe was studied by means of time-resolved photoluminescence spectroscopy. The dynamics showed a bi-exponential decay and the decay constants were measured as a function of number of layers and temperature.

In the last chapter of this **book** (Chapter 6) and the last part of my PhD, heterostructures of TMD monolayers were successfully fabricated and stud-ied. Some first interesting experimental results were already obtained, as for instance time-resolved PL spectra and second harmonic generation, but differ-ent experiments are planned to be performed. This was only the preliminary work for a broader perspective of experiments and projects. In the short-term, a free-electron laser experiment will be tried out with the goal of investigating the intraexcitonic transitions of interlayer excitons. Furthermore, an investigation of the role of hopping in the moiré periodic potential of interlayer excitons has already been started.

Moreover, an experiment on the effects of infrared radiation on few-layer InSe was already carried out at FELBE showing promising results. The successful InSe project and the developed technology necessary for the fabrication of vdW heterostructures lead to the natural idea of combining different materials for the fabrication of heterostructures, e.g. p-$MoSe_2$/n-InSe heterostructures.

A very appealing idea to pursue is to try to use and engineer van der Waals heterostructures for making two-dimensional materials active in the infrared region by taking advantage of intersubband transitions. This topic, which has almost not been investigated yet, clearly has potential for technological applications and new exciting research.

A. Appendix

A.1. Exciton optical spectra

In perfect analogy with the hydrogen atom, the solutions of Wannier equation (Eq. (1.11)) have bound states and a continuum for $E_n < E_g$ and $E_n > E_g$, respectively. The bound state solutions are called exciton wavefunctions and are:

$$\psi_{n,l,m}(r) = C_{n,l,m}\left(\frac{2r}{na_0}\right)^l e^{-\frac{r}{na_0}} L_{n+l}^{2l+1}(r) Y_{l,m}(\theta, \phi), \tag{A.1}$$

where n, l, and m are the quantum numbers, L is a Laguerre polynomial, and Y is a spherical harmonic. These functions define the distribution probability of all the exciton states. A full derivation of the hydrogen problem can be found in many textbooks, for instance here [187].

In order to find the expressions for the optical response for the bound states, Equation (1.10) is used and the microscopic polarization is expressed in the basis of the exciton wavefunctions:

$$P_{vc}(,r,\omega) = \sum_n b_n \psi_n(r). \tag{A.2}$$

After some algebra, one gets an expression for the exciton susceptibility:

$$\chi(\omega) = -2d_{cv}^2 \sum_n |\psi_n(r=0)|^2 \left(\frac{1}{\hbar(\omega + i\gamma) - E_g - E_n} - \frac{1}{\hbar(\omega + i\gamma) + E_g + E_n} \right), \tag{A.3}$$

where the susceptibility is $\chi(\omega) = \frac{P(\omega)}{\epsilon_0 E(\omega)}$. The first addend of the last equation is the resonant part and gives a very high value for $\hbar\omega = E_g + E_n$, i.e. when the photon energy $\hbar\omega$ is in resonance with the exciton states. The second part is non-resonant and can be neglected for photon energy close enough to the band-gap energy.

Another important simplification is that the exciton wavefunctions at $r = 0$ are always vanishing except for $l = 0$ and $m = 0$. This means that the allowed transitions are only between the ground state and the s-states of the exciton. This is not true in the case that valence and conduction bands have even parity, e.g. excitons in copper oxide [188]. The fact that the wavefunctions have to be evaluated in the origin has the intuitive interpretation that the electron and the hole have to be in the same position in order to recombine.

The coefficients in Equation (A.3) are $|\psi_n(r = 0)|^2 = \frac{1}{\pi a_0^3 n^3}$. The weights of the exciton states decreases as n^{-3}.

Absorption is related to susceptibility by the following formula:

$$\alpha_X(\omega) = \frac{2\omega}{c} \operatorname{Im}(\chi(\omega)) \propto \sum_n n^{-3} \frac{\gamma\omega}{(\hbar\omega - E_g - E_n)^2 + \gamma^2}. \tag{A.4}$$

Therefore, the exciton states show up as a series of Lorentzian functions at energies $E_{n'} = E_g + E_n = E_g - \frac{E_0}{n^2}$ below the electron-hole band gap.

To get the full absorption spectrum of a semiconductor, the continuum spectrum has to be included in the total absorption, i.e. $\alpha(\omega) = \alpha_X + \alpha_B$, where α_B is the free electron-hole absorption. Skipping the calculation that is anyway analogous to the above calculation, one gets:

$$\alpha(\omega) \propto \sqrt{\Delta}\frac{\hbar\omega}{E_0}\left(\sum_n n^{-3}\frac{\gamma}{(\hbar\omega - E_g - E_n)^2 + \gamma^2} + \Theta(\Delta)\frac{\pi e^{\frac{\pi}{\Delta}}}{\sinh\left(\frac{\pi}{\sqrt{\Delta}}\right)}\right), \tag{A.5}$$

where $\Delta = \frac{\hbar\omega - E_g}{E_0}$, and Θ is the Heaviside function. This is called Elliott formula and it consists of a series of Lorentzians (exciton states) and a continuum above band gap that resembles the free particle picture of the density of states.

However, if the hole-electron interaction had been neglected, the absorption would have scaled as the square root of Δ. Here, the Coulomb interaction changes the scaling. In fact, for $\Delta \longrightarrow 0$, the second addend of Equation (A.5) tends to $\Delta^{-\frac{1}{2}}$. Therefore, the absorption becomes enhanced and without any frequency dependence. This factor is called Sommerfeld factor.

A.2. Fitting function for exponential decay

In order to get an accurate decay function for time-resolved experiments, the instrument response function has to be taken into account. For most of the experiments, a Gaussian function is a very decent approximation ($F(t) = \frac{1}{\sqrt{2\pi}\sigma} \exp\left(-\frac{t^2}{2\sigma^2}\right)$).

Assuming an exponential decay function of the ensemble that has been probed ($D(t) = N \exp\left(-\frac{t-t_0}{\tau}\right)\Theta_H(t - t_0)$), the correct fitting function is the convolution:

$$
\begin{aligned}
I(t) = D(t) * F(t) &= \\
&= \frac{N}{\sqrt{2\pi}\sigma} \int_{-\infty}^{\infty} e^{-\frac{y-t_0}{\tau}} \, \Theta_H(y - t_0) \, e^{-\frac{(y-t)^2}{2\sigma^2}} \, dy = \\
&= \frac{N}{\sqrt{2\pi}\sigma} \int_{t_0}^{\infty} e^{-\frac{y-t_0}{\tau}} \, e^{-\frac{(y-t)^2}{2\sigma^2}} \, dy \\
&= \frac{N}{2} e^{-\frac{t-t_0}{\tau} + \frac{\sigma^2}{2\tau^2}} \, \mathrm{erfc}\left(\frac{t_0 - t}{\sqrt{2}\sigma} + \frac{\sigma}{\sqrt{2}\tau}\right)
\end{aligned}
\tag{A.6}
$$

In order to get the time resolution of the setup from this fitting function, the relation between the full width at half maximum of the Gaussian function and the broadening σ has to be used, i.e.:

$$
\Delta t_{IRF} = 2\sqrt{2\ln(2)}\sigma \simeq 2.4 \cdot \sigma.
\tag{A.7}
$$

In many cases, the decay function is a multi-exponential function. The convolution is an associative operation, therefore fitting function of a multi-exponential decay is simply the sum of the convolution between the instrument response function and each exponential decay.

A.3. FEL pulse length

The FEL spectrum at $E = 32$ meV is shown in Figure A.1. In order to evaluate the pulse length, the spectrum is fitted with a Gaussian function ($f \sim e^{-\frac{(E-E_0)^2}{2\sigma^2}}$).

The standard deviation of the Gaussian is $\sigma = 0.14$ meV that corresponds to a full width at half maximum of $FWHM = 0.33$ meV. The pulse length is then

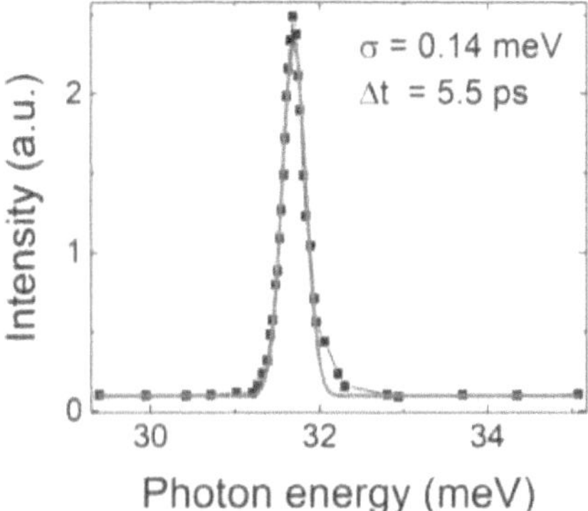

Figure A.1.: FEL spectrum at $E = 32$ meV. The estimated pulse length is 5.5 ps.

estimated with the uncertainty principle as:

$$\Delta t \simeq 4 \ln(2) \frac{\hbar}{FWHM} = 5.5 \ ps. \tag{A.8}$$

A.4. Theory of trion spectra

A trion is a quasiparticle observed in semiconductors composed by two electrons and one hole or two holes and one electron. Electrons and holes are bound together via Coulomb interaction. This three-body system admits a spin-degenerate bound state but, interestingly, only one, i.e. the ground state [152].

The aim of this section is to give an expression for the optical susceptibility of a direct band-gap semiconductor with parabolic dispersion. In particular, the three-body interaction is included in order to reach an analytic expression for the trion resonance. The approach developed by Esser *et al.* is taken as main reference [153, 155].

In a two-band model, the Hamiltonian of the system is given by $H = H_0 + H_{em}$

where

$$H_0 = \sum_k E_k^e c_k^\dagger c_k + \sum_k E_k^h d_k^\dagger d_k + \frac{1}{2} \sum_{k,k',q} v_q^{ee} c_{k-q}^\dagger c_{k'+q}^\dagger c_{k'} c_k -$$

$$\sum_{k,k',q} v_q^{eh} c_{k-q}^\dagger d_{k'+q}^\dagger d_{k'} c_k + \frac{1}{2} \sum_{k,k',q} v_q^{hh} d_{k-q}^\dagger d_{k'+q}^\dagger d_{k'} d_k, \tag{A.9}$$

where $E_k^{e,h}$ are the hole and electron band dispersion, $c^\dagger(d^\dagger)$ and $c(d)$ are the creation and annihilation operators for electrons (holes), and $v_q^{ee} = v^{eh} = v^{hh} = v_q = \frac{e^2}{\epsilon \Omega q}$ is the Coulomb potential in 2D. Ω is a normalization factor.

For simplicity, the background hole population is assumed to be vanishing, i.e. only the negative trion is considered. Therefore, the hole-hole interaction term will be assumed to be negligible with respect to the other terms for the rest of this discussion.

The Hamiltonian related to the external field is given by

$$H_{em} = \frac{eA(t)}{m_0} \sum_k p_k^{cv} c_k d_{-k} + h.c. \tag{A.10}$$

where p is the interband matrix element. The external electromagnetic field is given by the vector potential $A(t)$ can be assumed to be a δ-function in time, i.e. $A(t) = A_0 \delta(t)$.

In order to find the optical susceptibility, an expression for the time-dependent microscopic polarization is needed. Afterwards, by Fourier transform, the spectral response of the polarization is obtained.

In the density matrix formalism, the time evolution of an operator is given by the the Von Neumann equation $i\hbar \partial_t P_k = [H_0 + H_{em}, P_k]$. The microscopic polarization is defined as the observable $P_k(t) = \langle c_k^\dagger d_{-k}^\dagger \rangle$. Therefore, one gets:

$$(E_k^e + E_k^h + i\hbar \partial_t) P_k(t) - \sum_q v_q^{eh} P_{k-q}(t) - \sum_{q,k'} v_q^{ee} \langle c_{k-q}^\dagger c_{k'+q}^\dagger d_{-k}^\dagger c_{k'} \rangle$$

$$+ \sum_{q,k'} v_q^{eeh} \langle c_k^\dagger c_{k'-q}^\dagger d_{q-k}^\dagger c_{k'} \rangle = -\frac{eA(t)p_k^{cv}}{m_0}(1 - f_k^e), \tag{A.11}$$

where f_k^e is the electron distribution function. The hole distribution function is not included since zero hole background population was assumed. The terms

$\langle c^\dagger d^\dagger d^\dagger d \rangle$ are also neglected.

In Equation (A.11), the negative trion appears already as $T_{q,k,k'} = c_k^\dagger c_{k'}^\dagger d_{q-k-k'}^\dagger c_q$. In fact, the term creates a state with two electrons and a hole plus an empty space in the Fermi sea. To study the evolution of the trion polarization T, the Von Neumann equation is applied to this operator. Six-term operators will be obtained. Neglecting the quadratic terms in the electron number (e.g. $\langle c_1^\dagger c_2^\dagger c_3^\dagger d_4^\dagger c_5 c_6 \rangle \simeq 0$), the equation of motion for T is:

$$(E_{k_1}^e + E_{k_2}^e + E_{q-k_1-k_2}^h - E_q^e + i\hbar\partial_t)T_{q,k_1,k_2} + \sum_{q'} v_{q'}^{ee} T_{q,k_1-q',k_2+q'} -$$

$$- \sum_{q'} v_{q'}^{eh}\left(T_{q,k_1-q',k_2} + T_{q,k_1,k_2-q'}\right) = -\frac{eA(t)}{m_0} f_q^e \left(p_{k_1}\delta_{k_2,q} - p_{k_2}\delta_{k_1,q}\right). \tag{A.12}$$

From now on, only the singlet state of the trion polarization solutions will be considered, i.e. the antisymmetric solutions. The equation of motion of the trion and the exciton polarization will be resolved using the respective Green functions. And afterwards, expanding the Green functions in terms of the excitonic and trionic wavefunctions of the three-particle Schrödinger equation, an expression for polarization and susceptibility will be obtained.

Choosing the momenta $k_1 \longrightarrow k_1 + \frac{m_e}{M_T}$ and $k_2 \longrightarrow k_2 + \frac{m_e}{M_T}$, the equation of motion for the trion Green function is:

$$\left(\frac{\hbar^2(k_1^2 + k_2^2)}{2\mu} + \frac{\hbar^2 k_1 k_2}{m_h} - W_q + i\hbar\partial_t\right)G_{q,k_1,k_2}^T + \sum_{q'} v_{q'}^{ee} G_{q,k_1-q',k_2+q'}^T -$$

$$- \sum_{q'} v_{q'}^{eh}\left(G_{q,k_1-q',k_2}^T + G_{q,k_1,k_2-q'}^T\right) = \Omega\delta(t)\left(\delta_{k_2,\frac{M_X}{M_T}q} + \delta_{k_1,\frac{M_X}{M_T}q}\right), \tag{A.13}$$

where $\mu = (m_e^{-1} + m_h^{-1})^{-1}$ is the reduced exciton mass, $M_X = m_e + m_h$ is the exciton mass, $M_T = 2m_e + m_h$ is the trion mass, and $W_q = E_q^e - \frac{\hbar^2 q^2}{2M_T} = \frac{\hbar^2 q^2 M_X}{2m_e M_T}$ is an energy shift of the trion resonance due to excess of kinetic energy of the trion center of mass when the electron momentum is transferred to the trion. This term is called electron recoil. This term is positive, since the trion mass is always greater than the electron mass. This means that this term reduces the energy of the trion transition.

The trion and exciton Green functions can be expanded in terms of their respective wavefunctions in frequency space:

$$G^X(k, k', \omega) = \sum_n \frac{\phi_n(k)\phi_n^*(k')}{E_n^X - \hbar\omega} \tag{A.14}$$

$$G_q^T(k_1, k_2, \omega) = \frac{1}{\Omega} \sum_n \frac{\psi_n(k_1, k_2) \sum_{k'} \psi_n^*(k', q)}{E_n^T - W_q - \hbar\omega}, \tag{A.15}$$

where ϕ and ψ are the exciton and trion wavefunctions. The trion Green function admits also a continuum at energies higher than the exciton energy but this was not considered in the previous equation.

Therefore, combining the equations of motion for exciton (Eq. (A.11)) and trion (Eq. (A.13)) polarization, and using the relation $\chi(\omega) \propto \sum_k \frac{P_k(\omega)}{A_0}$, the total susceptibility becomes

$$\chi(\omega) = \chi_0 \left((1 - N_e) \sum_n \frac{\phi_n^2(0)}{E_n^X - \hbar\omega} + \frac{1}{\Omega} \sum_q f_q^e \sum_n \frac{\left| M^T(q) \right|^2}{E_n^T - W_q - \hbar\omega} \right), \tag{A.16}$$

where $M^T = \int d\rho \psi_n(\rho_1 = 0, \rho) e^{iq\rho \frac{M_X}{M_T}}$ gives the optical strength of the optical transition.

The exciton energies E_n^X can be calculated solving the hydrogen problem as discussed above (Section A.1). Analogously, the energy of the trion ground state can be estimated with the hydrogen anion problem. With a simple proportion, $E^T = E_{H^-} \cdot \frac{E^X}{E_H}$, where E_H and E_{H^-} are the energy levels of the hydrogen atom and hydrogen anion, respectively. In the case of MoSe$_2$ monolayer, the exciton binding energy is $E^X = 0.55$ eV [17]. The energy needed to ionize one electron in the hydrogen anion is $E_{H^-} = 0.75$ eV and the electronic ground state of the hydrogen atom is $E_H = 13.6$ eV [187]. Therefore, the trion binding energy is around $E^T = 0.03$ eV, i.e. in very good agreement with the experimental value [189].

The exciton and trion components to the susceptibility have a familiar form that reminds of the Elliott formula. It is important to notice that the exciton resonance energy is not affected by the free electron population. Differently, the trion energy is redshifted by the high-momentum electrons.

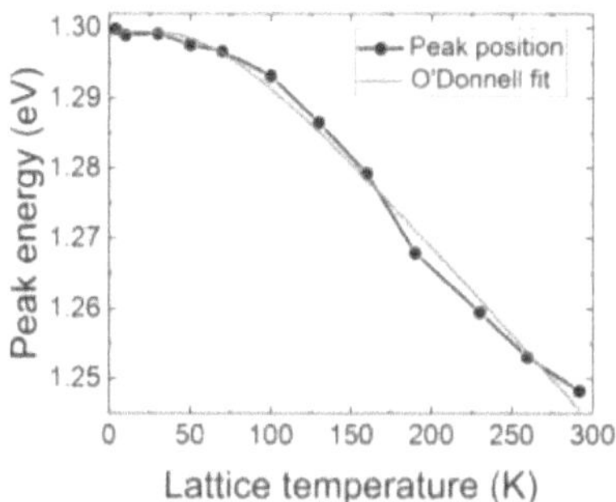

Figure A.2.: Energy band gap of InSe obtained from the model discussed in the main text (Equation (3.1)) with respect to lattice temperature. The fit curve in red has been produced with the O'Donnell model.

A.5. Energy band-gap shift with temperature in InSe

Figure A.2 shows the temperature dependence of the energy band gap of a 24-layer thick InSe sample. A typical redshift with temperature of the PL emission is observed as consequence of the decrease of the band-gap energy due to the lattice expansion and to the interaction with phonons. The semi-empirical O'Donnell model $(E_g(T) = E_0 - S\ \hbar\omega(\coth(\frac{\hbar\omega}{kT}) - 1))$ [190] can reproduce the dependence of the peak energy position E_g on temperature with good accuracy. From the fitting, $E_0 = 1.299$ eV is the PL energy at zero Kelvin, $S = 1.55$ is a dimensionless coupling constant, and $\hbar\omega = 18 \pm 1$ meV $= 145 \pm 8$ cm^{-1} is an average phonon energy. The average phonon energy obtained by this fit is close to the values of the energy of the optical phonons as measured by Raman (see Figure A.3) and FTIR spectroscopy [191–193].

In Figure A.3, a Raman spectrum of the 24-layer InSe sample is shown. Three phonon resonances are observed at 115 cm^{-1}, 177 cm^{-1} and 227 cm^{-1} in good agreement with literature [191]. The observed peaks are slightly redshifted with respect to literature values of samples with the same thickness. This fact could indicate a slight strain in the flake [194].

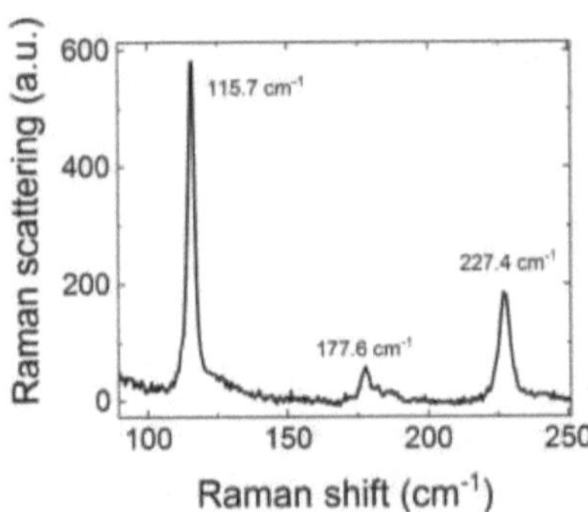

Figure A.3.: Raman spectrum of thin-layer InSe. The excitation wavelength is 532 nm. The thickness of the flake is 20 nm.

A.6. Decomposition of the PL decay

Figure A.4(a) shows the PL emission as a function of time and photon energy of a 24-layer InSe crystal. First, the PL data were integrated with respect to the photon energy and then the obtained PL decay was fitted with a bi-exponential function.

After that, the time constants ($\tau_1 = 7.7 \pm 0.2$ ns and $\tau_2 = 49 \pm 6$ ns in the case shown in figure A.4) were fixed and the PL decay was fitted with a bi-exponential function at each photon energy. Only the weights of the two components were adjusted at each photon energy. Figure A.4(b) shows the obtained fit of the PL data. Figures A.4(c) and A.4(d) show the decomposed fast and slow components of the PL decay.

The two components are spectrally separated now. Figure 5.9 in Chapter 5 shows the spectrum of each component separately.

The same procedure was applied to samples with other thicknesses with similar results.

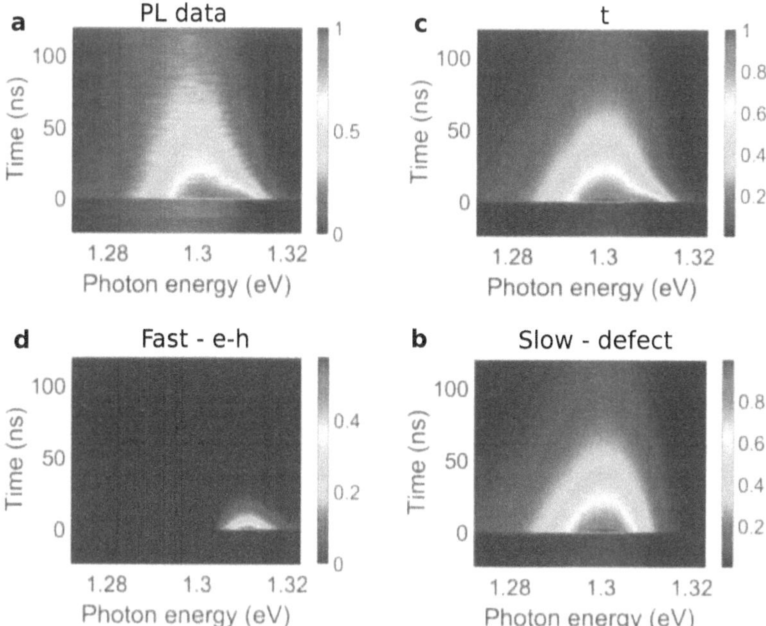

Figure A.4.: (a) Normalized PL of a 24-layer thick sample as a function of time and photon energy. (b) Bi-exponential fit of the PL at each photon energy. The decay constants were kept fixed for each photon energy ($\tau_1 = 7.7 \pm 0.2$ ns and $\tau_2 = 49 \pm 6$ ns). (c) and (d) Fast and slow components of the PL emission as extracted from the fit.